21st Century Poultry Breeding

by

Grant Brereton

Edited by Sara Roadnight

Published by Gold Cockerel Books

ISBN 97809 47870 577

Contents

Chapter 3

A look at the genes:

FOREWORD

It is a great pleasure to write a foreword for this book. I have known Grant Brereton for many years and have witnessed his activities with much interest. While many enthusiasts, myself included, have been content to breed birds solely for exhibition, striving to achieve perfection against a given standard, Grant has gone beyond this. Although playing a part in the UK poultry show scene, he has been driven primarily by a fascination for how different plumage colours and patterns are made up, and by an urge to investigate exactly what can be achieved genetically. He has gained an insight and understanding of poultry plumage genetics that is both remarkable and enviable; his willingness in this volume to share his knowledge, explaining the genetic mechanisms for the exhibitor and breeder in a simple and lucid manner, is most laudable.

Grant first became interested in poultry at the age of six, and has kept fowl more or less continuously since then. The son of Alan and June Brereton, he was schooled in Denbigh, North Wales and went on to gain a BTEC diploma in Business and Finance at Kelsterton College, Deeside. He has since been a town planner and run his own businesses.

Grant became intrigued by poultry genetics when at the age of 21, his girlfriend gave him a Christmas present of a pair of Buff Laced (which could more logically be called "White Laced Red") Wyandottes, which she had purchased from the late Gilbert May of Cornwall. The offspring produced from the pair perplexed him greatly. Since they were purebred standard birds, he expected all the progeny to be Buff Laced like their parents. However, only approximately half were Buff Laced, the rest were about equally divided amongst Gold Laced (i.e. "Black Laced Red"), and birds that were almost totally White. Fascinated by this failure of a pure breed to breed true, he became determined to understand what was going on. All became obvious when he read about Mendel's Law and the mechanism of inheritance. Some time later, he was very fortunate to meet the poultry geneticists, the late Dr Clive Carefoot of the UK and Brian Reeder of the USA. From then on, his enthusiastic interest in the subject has continued to blossom and by now he is the leading authority on the genetics of domestic fowl among the UK poultry fancy.

For more than a decade now, Grant has studied fowl genetics from his own observations, from word of mouth, and from elongated emails over many years with Brian Reeder. He has applied his knowledge to the creation of many new Wyandotte colours, including (to name but a few) the Pyle, Buff-Columbian, White Buff-Columbian, Blue-Barred, Chocolate-Partridge, and Wheaten. He has also generously advised many breeders and exhibitors, and assisted in their various breeding projects. During the last 11 years he has conducted countless test matings to determine the genetic make-up of purebred strains. He rescued the Large Silver Pencilled Wyandottes from the point of extinction, and has since shown them with great success.
Over the years, Grant has kept most of the UK breeds, and has written many articles for UK poultry magazines, including Practical Poultry, Smallholder, Country Smallholding, and Fancy Fowl. He has also written for foreign magazines, including Avicultura in the Netherlands, Gelfluegel Boerse in Germany, Hobby Farms in the USA, and the Australian Wyandotte Club.
Most importantly, in 2007 he became the editor of one of the leading UK poultry publications, Fancy Fowl, from which time he has been well known to all fanciers. In this post he has shown flair, enthusiasm, and dedication; the publication has gone from strength to strength and continues to do so.
This book is a vital part of the reading and reference material of every poultry fancier. It is hard to imagine anyone who will not find something in it to fascinate them, whether it is the genetic details of their own breed, or the broad outline of how poultry plumage genetics operates to generate the wonderful permutations of patterns and colours that we see in the domestic fowl.

Geoff Parker

Past President of the Poultry Club of Great Britain
President of the Partridge and Pencilled Wyandotte Club
Derby Professor of Zoology, University of Liverpool

INTRODUCTION

We are approaching a new age in poultry breeding – at the time of this writing, celebrity chefs and the like are exposing welfare conditions of egg laying and poultry raised for meat which has led to an upsurge in the number of people wanting to "do it themselves."

As we strive to become more self sufficient, we can look at the available options with poultry. There are many breeds available to suit our individual needs and this book looks at some of the more popular "pure" breeds and offers a basic guide to methods of sourcing quality stock.

Many people start off with a few hens for the garden, which in turn gives them fresh, free-range eggs. However, there is a natural progression, and many "backyard" poultry keepers end up showing their birds at regional, and sometimes national poultry shows. It is for this reason that I have included a chapter on how to breed in numbers and select better show quality specimens.

Over the last ten years of "breeding seriously" I have conducted many experiments and learnt a lot about how the genes for plumage operate, and how they interact with each other to produce all the wonderful colours and patterns seen in poultry today.

A basic knowledge of the genes is invaluable in any breeding programme and by applying the laws of Geneticist, Gregor Mendel (Mendel's Law) it is possible, in most cases to determine the quantities of desirable or undesirable traits in the offspring we produce from our chickens. For example, "Sex linkage" is invaluable when one only intends rearing females.

Crossing a genetically Gold male over genetically Silver females (i.e. a Rhode Island Red male over Light Sussex females), results in brown downed female chicks and yellow downed male chicks – this is just one example of "Sex linkage" which I will cover in more detail.

Whether you are looking for something to get you started, or you are seeking more knowledge of how the plumage patterns work, this book aims to be "dual purpose" and serve as a general guide to poultry breeding combined with a section to make understanding the genetics that little bit easier.

Grant Brereton, Ruthin, 2008

WHAT IS A BREED?

In reality, the word "breed" is a word that best describes a desirable group of genes or mutations that we as humans have selected for over time – whether it be in dogs, cats, cattle, sheep or poultry. My belief is "breeds" don't really exist without human intervention, whereas "species" quite clearly do.
The obvious difference between a breed and a species is that "species" breed true in the wild, for example, the Red Jungle Fowl, which is assumed by many geneticists to be the original ancestor of all poultry breeds.
However, if we look at the other three varieties of Jungle Fowl, the Ceylon, the Green and the Grey (Sonnerats), and their ability to produce fertile or semi fertile offspring when crossed with domestic fowl and sometimes each other, then just maybe, the age old assumption that all fowl derived from the Red Jungle fowl (first concluded by Darwin), needs to be re-evaluated. It was the Geneticist, Hutt who first challenged the theory.
At the time of this writing, many Geneticists are coming round to the idea that not all fowl derived from the Red Jungle Fowl and are exploring the aforementioned prospect that other species of Jungle Fowl played a part in paving the way for domestication as we know it today.
In essence, recent exhaustive studies have concluded that the yellow skin gene in fowl derived from the Grey Jungle Fowl.
Perhaps the most successful of Jungle Fowl species to be crossed to domestic fowl is the Grey Jungle Fowl. These hybrids have been named as "Bengals" and are extremely fertile, proving to be capable of crossing either as sibling matings or back to either of their respective parental groups, the Grey Jungle Fowl, or domestic fowl.

The above photo sequence shows a Black-Red Phoenix male (left), a Grey Jungle fowl female (middle) and a resultant "Bengal" offspring (right).

Grey Jungle Fowl (above left) will hybridise with the Red Jungle Fowl (above right) in the wild, on the rare occasions that their territories cross.

Above left and bottom right, the result of a cross between the Green Jungle Fowl middle right and Silver Phoenix. These Hybrids have a chromosomal mismatch, so backcrossing to the Green Jungle Fowl is unlikely, however the hybrid males can be successfully crossed to domestic game hens.

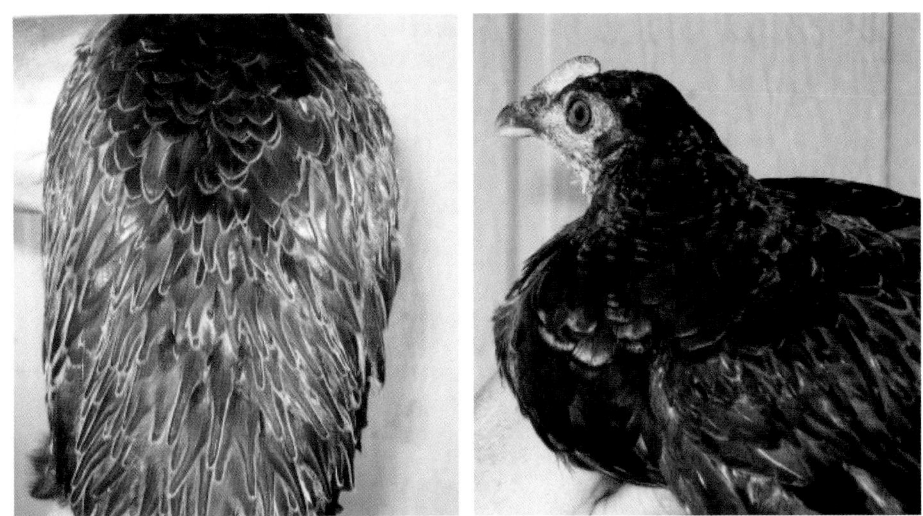

Above: a Green Jungle Fowl male, displaying great plausibility for the origin of the Blackening genes in poultry.

A pair of Red Jungle Fowl.

Whether or not the new thinking is accepted by the majority will likely be a subject to cause conjecture amongst those with an interest, for many years to come. More importantly though, for the quantifying of genes and determination of how plumage patterns work and are related to each other, it is essential in "Mendel" terms to have a "wild type" which is the basis for applying the principles of Mendel's Law. This will be covered in more detail later. So far and until proven otherwise, the "wild type" plumage will remain to be considered that of the Red Jungle Fowl (pictured above).

Breed Standards

In many countries where the exhibition of pure breed poultry is taken seriously, there is a publication of "standards", usually published by the main governing poultry body. In the UK, it is the Poultry Club of Great Britain. These standards were set out so that the characteristics and defining features of all breeds are abundantly clear, and so that breeders have a set of rules to follow in their quest to produce excellent quality show stock. In America, the standards book is entitled "The Standard of Perfection". In Australia, the standards book is entitled "The Australian Poultry Standards".
Without "Breed Standards" there would be no such thing as "Poultry Breeds" because everyone would have their own interpretation of how their particular breed should appear. Though the current "standards" are debated on a small scale, in most cases they are clear on how the correct examples of each breed should represent themselves, and furthermore, usually include photos to reinforce their requirements.
Though each breed has its own set of standards, and includes the necessary rules for the varying standardised colours within that breed, there are common faults which should not be tolerated across the broad spectrum of poultry breeds. Such faults include a wry tail, bow legs, split wings (except some Game varieties) and so forth.

Generally, each book of standards is there to maintain the quality of all pure breeds, bred and shown within their jurisdiction and not just exhibition lines. However, not all countries would agree on certain standards, for example, the Leghorn fowl which originated in Italy. The breed standards for Leghorns in the UK and the US are quite different. So, it goes to prove that while most of the time standards are similar, each country has its own interpretation of what certain breeds should look like.

The standards of each breed included in the "British Poultry Standards" book were drawn up and submitted by each respective breed club affiliated to the Poultry Club of Great Britain.

In the "British Poultry Standards" the following attributes are described and a scale of points is drawn up which adds up to 100.

Each breed varies in how many points it assigns to each feature of the fowl. This allows judges to determine the scale of priority when judging. For example:

Colour:	25	Type:	20	
Head:	15	Size:	10	
Condition:	20	Legs:	10	100 Points

Standardising a new Breed or Colour of Poultry:

Each country is different, but here in the UK, the Poultry Club of Great Britain has a set of basic requirements which can be obtained on request for those in pursuit of standardising a new breed, or colour of an existing breed of poultry.

As time goes by, these requirements may change a little. However, I'm sure there will always be a requirement for minimum generations of official leg rung stock to be in the hands of at least a "required number" of breeders, with the ability to prove that the proposed birds "breed true". It will likely always be the case that a fully plausible and comprehensible set of standards have to be drawn up, and then vetted and submitted to the PCGB by the appropriate Breed Club to be considered for standardisation.

Sourcing Stock

There are many ways of sourcing poultry nowadays, particularly now we have the use of the internet. Ebay has been one method of buying and selling hatching eggs, with prices often soaring, even though the buyer has no guarantees of the eggs being fertile.
Since poultry keeping has become more popular, and named "the fastest growing hobby," many newcomers have emerged with fancy looking business names and websites claiming to be "specialist breeders" of a range of poultry breeds, even though they are fairly new to the hobby of poultry keeping.
In my view, there is nothing wrong with such websites or breeders as they are usually very meticulous people who rear their birds with great care. If someone is prepared to take the time to create a website proudly displaying their stock, the chances are their birds will be half decent. Often such businesses, or hobbyists with surplus stock, allow visitors and are very friendly people.
Before you purchase any stock, it is essential to determine in your own mind exactly what you want - whether you are seeking an egg laying breed, something attractive for the garden, or a mixture of the two. The vast choice of breeds, patterns, colours, and features in poultry makes it a very difficult decision to reach, and sometimes the worst thing you can do is to visit a "Rare Breed Poultry Sale" as this only spoils you further for choice.
In my view, by far the best method of sourcing stock is to locate the relevant breed club through one of the listings in poultry magazines. This way the secretary can point you in the direction of breeders nearest to you. Breeders don't always produce superstar show stock and often have surplus stock which is quite reasonable, so it would be wrong to assume that such breeders only have show quality birds for sale.
Whether you are seeking Hybrid stock or Pure Poultry, the following options are available to you:

A/ Contacting the relevant Breed Club
B/ For Sale listings in the back of magazines
C/ Rare Breed Poultry Auctions
D/ Online breeders advertising on forums
E/ Adverts in your local Newspaper
F/ Word of mouth

G/ Adverts in your local Feed Merchant
H/ Rare Breed Centres
I/ The Rare Breed Survival Trust

Light Sussex, large fowl.

Welsummers, large fowl.

Breeding in Numbers

Whether you are an experienced poultry breeder or relatively new to the hobby, there are some common factors which have to be learnt in order to improve your stock. The principle factor in breeding is the "art of selection". This basically means acquiring the knowledge to understand what is desirable and what is not desirable when assessing the stock you will keep and what will be moved on to the less "picky" among us. For example, garden poultry keepers are more than happy to pay a little less (or quite considerably less) for a reasonable example of a breed which is not quite good enough for the show pen.

The best way to learn just what constitutes a good example of your chosen breed is to visit shows, talk to breeders and learn about type. Investing in a copy of the relevant poultry standards is a wise thing to do, because it will enable you to learn not only what features are desirable in your breed, but just as important, which characteristics are not.

A common misconception for many newcomers is that a particular breeding trio will replicate itself almost perfectly in the offspring it produces. Whilst this is a nice thought, it is very rarely the case and after years of personal experience and hours of discussion with breeders, it seems as though the art of "selection" is the key to not only maintaining the quality of your stock, but perhaps more importantly, improving it.

Breeds remain the way they do because of a few very skilled breeders who are dedicated to the cause. For them, it's a passion and some breeders have spent their whole lives maintaining certain strains, and in turn have been awarded the red cards and silverware at shows throughout the country.

Breeds can soon lose their signature attributes if no thought is given to the breed standard. To give an example, the rare Large Blue-Laced Wyandottes had so much emphasis placed on markings, that almost no consideration had been given to type, and back in 2001, the few remaining examples were almost "Leghorn shape". It was necessary to import birds from Germany to help rectify this. The moral of this story is that almost every breed could end up this way if it weren't for the few who keep the breeds as they are with their constant selection for desirable features.

I could never really understand the concept of breeding in numbers, (which basically means hatching more than just 4 or 5 chicks), until I realised that hatching such a small number of livestock would likely yield nothing of satisfactory value; in order to stand a chance of producing something as good, if not better than the parent stock, I would have to seriously consider hatching at least 30 to 40 chicks from each breeding pen.

The longer you breed poultry, and especially if you show, the more particular you become about trying to breed the "perfect bird". Though it doesn't exist, because no specimen is infallible, it doesn't stop breeders aiming as high as possible in the hope of producing that perfect show bird one day. My personal yardstick is, if I breed 20 chicks from a fairly good breeding pen, then the chances are that 5 will be desirable for the show bench, 10 will be average and 5 will be below average. I have found this ratio to be applicable to many breeds and when I use the term "below average" I refer to birds that are still a respectable representation of their breed, but would suit a garden keeper as opposed to a serious exhibiter.

Spangled Hamburgh pullets.

White Wyandotte males.

Light Sussex pullets.

Black Orpington males.

As the photos on the opposite page demonstrate, it is important to have a healthy choice from the offspring you produce. Not only do breeds share the same faults across the spectrum (split wings, bent toes, wry tails, duck footedness etc) which are selected against by breeders, they also encounter faults which deviate from the desired characteristics of each respective breed. To give an example, breeders of single combed breeds need to consider the amount of serrations, or "comb spikes" that their stock displays, whereas rose combed breeds don't share this concern. However, they have their own issues and depending on the breed, the way the leader (the back spike) points is a major factor when determining whether the bird in question will make the grade.

The good news is that "like" does produce "like" and as breeders, by far our best assets are our eyes. Granted, we can't select against the hidden faults in a line, but they do come to the surface eventually and as long as they are selected against in future breeding programmes, there is less chance they will recur. When selecting breeding pens, major consideration should be given to gauging what each bird will produce, for example, there is no use breeding from an Orpington female if she displays 10 comb serrations, when ideally, you are aiming for 5 or 6. That would not be logical unless the bird in question excelled in every other way and the plan was to breed such faults out in future.

Breeding is largely about having a vision in your mind of exactly what you're aiming for; this will of course go hand in hand with the poultry book of standards relevant to each country, that is, assuming you are wishing to breed pure-bred poultry. Breeding in numbers and allowing yourself selection from the offspring your line produces can be very gratifying: when you encounter some superstar birds, you will feel that it's all worthwhile.

Selection should be an easy way of improving a strain, especially considering what can be done with it. For example, one project of mine was to breed a Large Pyle Wyandotte. The best I could do was to breed a Pyle Leghorn to a line of unpatterned Partridge Wyandottes and sibling mate them and then breed back to the Partridge Wyandotte line for enough generations until my vision became more like reality. It was an enjoyable although lengthy task, but it was worth the wait in the end. If selection can turn one breed into another, then it certainly can be the tool which immensely improves one's strain, and I think is a very exciting prospect and key element to breeding.

The following photo sequence from A to F shows the bird produced from the first cross (Leghorn x Wyandotte) and the subsequent generations which eventually led to Pyle Wyandottes. I hope this demonstrates, if nothing else, just what can be achieved with a little imagination and determination....

Note that while the bottom right bird (pic F) is a far cry from the one in the top left photo (pic A), at this point, I still had to refine the shape before I could call it a "Wyandotte" and as a result, I successfully crossed this male (pic F) with a Partridge Wyandotte hen. This yielded most satisfactory results. The top left photo (pic A) is of the first cross between a Pyle Leghorn male and a Partridge Wyandotte female and the sequence of progressive males goes from left to right.

The Pyle Wyandotte

The above Pyle breeding pair (pics A & B), tested pure for the Wild Type genes by breeding them to pure Partridge Wyandottes. All males and females produced were either as the Pyle (pics A & B), or as pics C & D, proving that the only gene segregating (not pure) was the Dominant White gene I (standing for "Inhibitor of Black").

An Additional Pyle pair (pic E) were bred together to test not only the purity of the Wild type genes, but also to see whether as suspected, some virtually White offspring would be produced as a result of purity of the Dominant White gene. This proved to be the case and roughly 25% of the offspring were self White.
In addition to the expected percentage of Wild type offspring (Salmon Breasted females), in Pyle (pic F) and Black-Red (right bird in pic H), I also experienced 25% Partridge offspring (left bird pic H) with half of them carrying a single copy of the Dominant White gene, so in effect they were "White-Partridge" (pic G). Males as expected were either Pyle or Black-Red, so it was obvious that in order to purify the wild type genes and breed out the Partridge genes, selection could only really be carried out by retaining and breeding to the Salmon breasted females which were Dominant in appearance over the Partridge genes.

Improving a Strain

I'm a firm believer in the notion "If it's not broke, then don't try to fix it". There is little sense in crossing together the strains of two separate breeders just to call the strain your own. However, sometimes strains become so inbred that they reach the point of exhaustion and cannot procreate very well. It is likely that the few birds produced will be excellent visually, but other qualities such as egg production, disease resistance and general hardiness are lowered greatly.

Some examples are Professor Geoff Parker's strains of Partridge and Silver Pencilled Wyandotte Bantams. Geoff is a very well respected exhibitor and served a three year term as the president of the Poultry Club of Great Britain. His wins at the Major shows with his Partridge and Pencilled Wyandotte bantams prove beyond doubt that he is one of the most successful British breeders of Partridge and Pencilled Wyandottes. Geoff was a professor of Zoology at Liverpool University, so his understanding of genetics is very good. Geoff knew that Partridge and Silver Pencilled Wyandottes were seperated only by a single gene, this being "Silver".

Basically, a Silver Pencilled Wyandotte is a Partridge Wyandotte with the addition of the Silver gene. With this knowledge, Geoff crossed his two varieties together. This worked very well and he achieved outstanding success with them for many years, including his 1997 "Supreme Champion" Partridge winning pullet at the National Poultry Show. He would often breed his two varieties from the same pen.

However, it was in the year 2000 that Geoff arrived at a point when after hatching only 9 chicks out of 300 eggs, he realised it was time for a much needed "outcross" to inject some hybrid vigour and to revitalise his strain. He knew it wouldn't be easy, but after locating a Partridge Wyandotte male, he was able to cross it to his lines and this was carried out with great success. However, the important thing to remember here is that outcrossing is just the beginning. It's almost like starting over. Geoff knew that in order to resemble his strain, the progeny, and the subsequent progeny from the outcross would have to be bred back to his line. This process was carried out for 4 generations before Geoff was happy with his newly created lines of Partridge and Pencilled Wyandotte Bantams.

The moral of the story is to outcross only when it is absolutely essential to do so. If you purchase a trio from a market you have no idea of how closely related they are. It is likely they are brother and sisters. Some vendors put signs on their cages with the words: "Unrelated Trio". Whilst they may genuinely believe that to be the case, they can't possibly know for certain and just because they obtain a male from another breeder, without historic records, it is impossible to be sure. Even if the bird in question is imported from another country, it doesn't mean that the same strain wasn't exported there a few years previously.

When buying poultry, it is always ideal to view the breeding stock first, but it is not always possible to do so. It's not all bad though; if you buy a trio from an auction and they produce most satisfactory results, there is no need to question it. The mistake many people make is to rush to breed from the progeny of such trios in the quest to make use of them. They often pair birds that they know to be brother and sister, even though the parents of such birds were likely brother and sister to begin with. My yardstick is that if a trio breeds well, and produces good results, then I will keep breeding from that trio until it is too old to continue procreating. By this time, I will have made a cross or two from the progeny to the original parents in the hope of producing a trio much the same as the originals, and the process would continue.

If I encountered the beginning of faults from the above crosses and it were not possible to produce a trio that would breed much the same as the originals, then and only then, would I consider bringing in an outcross. The outcross would be selected with great care and sourced from the best breeder of the particular variety I was breeding. As with Geoff Parker, the simple rule of thumb would be to breed the progeny and subsequent generations back to the "superior parent" in the hope of replicating it, whilst at the same time keeping the perils of inbreeding at bay.

Geoff Parker's invigorated Partridge breeding pen.

Geoff Parker's invigorated Silver Pencilled breeding pen.

Hatching Large Fowl Early

Although it may be difficult to comprehend, the time of year a chick is hatched impacts greatly on the eventual size it will grow to. This was brought home to me in 1998, when my "May-Hatched" Large Partridge Wyandottes were dwarfed by the competition in the pens at the National Poultry Show. After speaking to breeders, not only of Wyandottes, but many varieties of Large Fowl, it became clear that it was necessary to hatch in January or February if I was to compete seriously in such classes.

Of course, the initial obstacle you come across when trying to hatch so early in the season is the lack of production - both on the male and female part. The solution is to extend the natural daylight in the winter months by adding artificial light. Although not ideal, this practice works very well and usually all that's required is a light plugged into a timer on a wall socket. For best results, it is advisable to begin by increasing the day length an hour at a time for the first few days and build it up that way. My timers came on between 6-9am and then 3-9pm, meaning that with the inclusion of natural daylight between the artificial light, the overall day amounted to 15 hours.

You will soon notice the increase in egg production and the cock in the pen will step to the fore in his role. He will be more virile and vocal than is expected for that time of year. However, the results will be worth it. It is important to feed well at this point, by using good quality "Breeders Pellets"; these can be sourced from most feed suppliers usually at a slightly higher cost than the Layers pellets. It is also important to feed greenstuffs at this time, and adding a teaspoon of cider vinegar per litre of water is very beneficial: this aids vitality and promotes a healthy appetite and good production.

Poultry Legends, Allan and Dinah Procter with a White Wyandotte Bantam male.

White Wyandotte males from a 40 year old strain, hatched in January.

White Wyandotte pullets, large and bantam.

A breeding pen of large Barred Wyandottes in January.

Some Barred pullets for selection.

Silver Pencilled pullets - like peas in a pod.

Silver Pencilled males.

Buff Wyandotte pullets - plenty of selection.

5 Partridge pullets selected out of a possible 20 for their breeding qualities.

Large Blue Laced Wyandotte males.

Hard Feather or Soft Feather?

When I was 8 years old, I acquired my first pure breed poultry, these were a trio of Large Light Sussex. When visiting a friend, and telling him my good news, his father said: "So then, are these new birds hard feather or soft feather?" Completely perplexed by his question, I just shrugged my shoulders. It wasn't until he asked what breed they were, and I answered Light Sussex that he explained to me the difference between hard feather and soft feather and that my birds were soft feather.

The difference between hard and soft feathered breeds becomes apparent when visiting a poultry show. The soft feathered varieties tend to be self explanatory, such as the Wyandotte, Orpington, Plymouth Rock and Faverolles, to name but a few. The hard feathered breeds tend to be all the Game varieties and perhaps obviously, Asian Hard Feather. It is easy to observe how much more "tightly" feathered they are, in particular some of the scantily feathered Asians.

Hard feathered varieties such as the Ko Shamo have neck hackle that ends before the bottom of the neck and should barely be any longer or fuller than the non-hackle neck feather. The feather on the body is very tight, hard and sparse, leaving bare red skin at the breast, vent and point of wing. It is the belief of many Asian Breeders that such hardness and sparseness of feather almost invariably results in split wing (where there is a distinct gap between the primary and secondary feathers when the wing is opened). This would be considered a fault in soft feathered breeds but is never a fault in breeds such as the Ko Shamo.

The plumage of hard feathered breeds should not be washed before a show as this would soften and loosen the feather; usually all that is needed is to wash the legs and feet, making sure that the backside is clean, and dressing the face with a spot of oil. A wipe with a silk cloth can add an extra gleam to the feather, but the major show attribute should be the presence of the bird itself. It can be quite an experience to see such a bird at a show for the first time.

As mentioned above, washing birds can soften and loosen the plumage, how much, is highly dependent on the additives to the water that each bird is washed in (Glycerine is renowned for having this affect). Breeders of soft feathered varieties such as the Orpington, Pekin and Cochin require very loose feathered "fluffy" birds, so the loosening of the feathers is something they embrace. They certainly breed for as much fluff and loose feathers as possible. The Orpington is probably the best example.

A Gold Dutch Bantam cockerel. This is classed as 'soft feather'.

A Ko-Shamo male. This breed is without a doubt 'hard feather'.

A Minorca hen. This is classed as 'soft feather'.

A Black-Red Old English Game male. This is classed as 'hard feather'.

A rearing pen of Black Orpington males. This breed is quite obviously 'soft feather'.

Old English Game, undubbed. This Breed is 'hard feather'.

Large Fowl, New Hampshire Red, Classed as 'soft feather'.

Double Mating - Partridge Wyandottes

The above photo and drawing depict Partridge Wyandottes in their true exhibition form (as they are standardised, and as they should be). The photos below show the breeding pens required to breed such specimens, respectively in male and female.

A pullet breeding Partridge Wyandotte pen = Exhibition Partridge female.

A cock breeding Partridge Wyandotte pen = Exhibition Partridge male.

A pullet Breeding male.

A cock Breeding female.

The above photos show the breeding partners, the "tools" from which the exhibition males and females are bred. The male pictured to the left is required to produce wonderfully patterned Partridge Exhibition females, and the female pictured right is used to produce Lemon Hackled, Black Breasted Exhibition males. Note her Lemon Hackle, coarse pencilling, and darker neck hackle. The true terminology would be left: pullet Breeding male, right: cock Breeding female.

The term "Double mating" means that exhibition males and females cannot be produced from the same breeding pen. An example is the Partridge Wyandotte, where exhibition males are required to display Lemon hackles and solid Black breasts (to name a couple of requirements), but these males have to be paired with females of their true genetic counterpart. These females are often very "mossy pencilled" birds and in no way resemble exhibition Partridge females, which display beautiful concentric black rings on beige feathers.

Just as the exhibition male needs its correct genetic breeding partner, exhibition females need to be paired with males which often display very speckled breasts and a completely different shade of hackle, which is Orange rather than Lemon.

Although "double mating" was often necessary when standards were drawn up (to achieve the correct coloured males and females respectively), times change and often classes are put on at the major shows to allow the "breeding

partners" to be exhibited.
Some may question the logic in this, but to give an example, not many breeders can recall Large Lemon Hackled Partridge Males in the UK (the correct exhibition type), so if it wasn't possible to exhibit "pullet breeding" males (the males required to produce Partridge females with wonderful markings), then no Large Partridge males would ever be shown, simply because the true "Exhibition" lines have been lost.
There are very few exhibition lines of Partridge Wyandotte Bantam Exhibition male "cock breeders" left in existence and good specimens are in the hands of a limited number of dedicated breeders.
Good examples of Exhibition Partridge males should display a "robin's breast" shade of orange on the shoulder, complimented by a solid Black breast and thighs with Lemon neck and saddle hackle surrounding solid Black striping.
Although I am fairly new to keeping Exhibition Partridge males, I have for a long time been aware of how they are said to breed. A Lemon Hackled breeding pen often produces 50% Lemon Hackled birds, 25% Orange Hackled birds, and 25% completely White birds. It was Dr Carefoot who first realised that a single copy of the recessive White gene was causing the Lemon Hackles in exhibition male lines.
Speaking to breeders, I find that many experience the same ratios as Dr Carefoot, even as far afield as Australia, whereas other breeders have never encountered white birds in their strains.
It would appear that the once limited strains of large exhibition Partridge "cock breeders" are now extinct in the UK and that not many fanciers can remember them. There are 2 photos in the late Ian Kay's book "Stairway to the Breeds" which depict winners from the mid thirties; one was owned by James Mellor and the other by a Mr T Maskrey.
With Dr Carefoot's findings, many people have set about recreating a line of Large Partridge "cock breeders", however, none has come into fruition to date. It was one of Dr Carefoot's last projects but he was unable, through ill health, to see it through.
I have regularly been asked, and more often in recent times, if it is even genetically possible to recreate a line of Large Exhibition "cock breeders". My thoughts are that anything is possible, but it would take at least four years of dedicated breeding and a lot of waste along the way. Maybe it's a

project for the future. It would be good to see the return of large Exhibition Partridge males.

Self Black Wyandottes have to be double mated in order to produce Black birds with yellow legs. The exhibition female line has superb yellow legs, however, the males required to produce such lustrous "beetle green sheen" yellow legged Black females, generally have a White under-colour, which renders them undesirable for the show bench.

Exhibition Black Wyandotte males require a Black under-colour and to this end, are paired with females which display dark legs. Little research has been carried out to determine just why this is the case. It is likely the result of impurity of genes such as Birchen in the genetic background of such males.

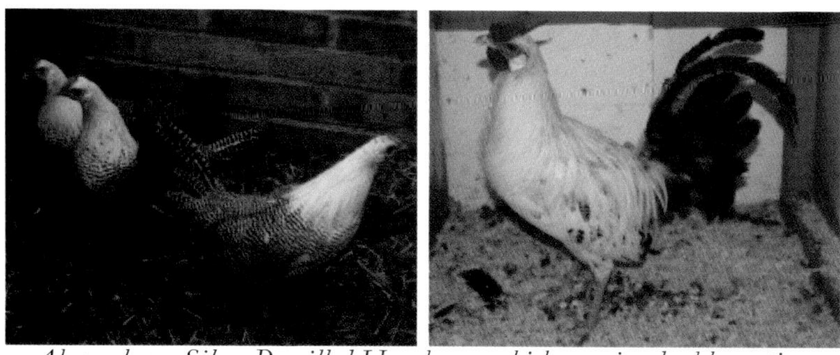

Above: large Silver Pencilled Hamburgs which require double mating.

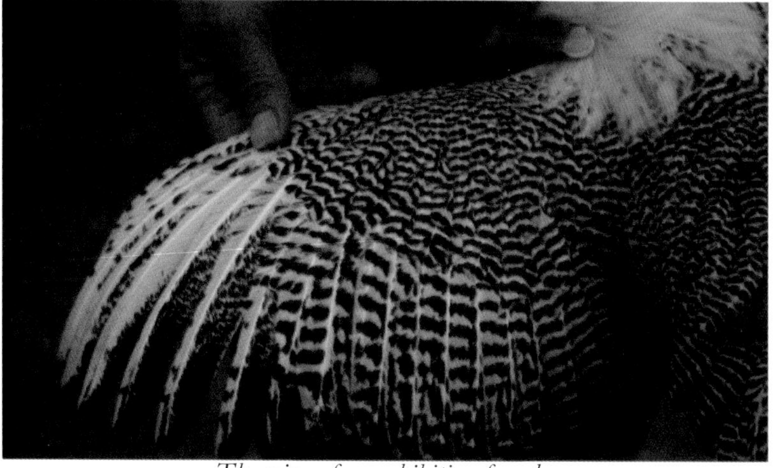

The wing of a exhibition female.

SOME POPULAR BREEDS

The Orpington

The Orpington Fowl comes in a vast range of colours, both Large Fowl and Bantam but predominantly Large. The original colour of Orpington was the Black, created by William Cook of Orpington, Kent, who first developed the variety in 1886. William Cook also developed the familiar Buff Orpington which is still a popular and attractive colour today. The original Orpington Club formed in 1887 to promote the Blacks and a Buff Orpington Club formed eleven years later, in 1898 to cater for the Buffs.

The Orpington is classed as a soft feather, heavy breed. Its defining characteristic is undoubtedly the abundance of fluff which any good specimen should display. It is required to have white legs (except the Black and Blue versions) and it should display a short, evenly spiked comb.

At the time of this writing, it is still permissible to exhibit Orpingtons which display a "Rose comb", however, this option will almost certainly be removed from the British Orpington Standard at some point, as the Rose comb Orpingtons have never been popular, even though the original White Orpingtons had Rose combs.

The main standardised colours are Black, Buff, Blue and White. However, Dr Clive Carefoot created a Chocolate Orpington, which I continued together with breeders, Rob Boyd and Richard Davies. Other non standard colours include Gold Laced, Crele, Red, Cuckoo, Erminette, and Jubilee.

White Orpington pullets.

Gold Laced Orpington trio.

Buff Orpingtons.

Blue Orpingtons.

Crele Orpington female.

Crele Orpington male.

Buff Orpington male.

Black Orpington female.

Buff Orpington pullets.

Superbly shaped Black Orpington male.

A pen of Buff Orpington Bantams just coming into feather.

Large Blue Orpington pullet.

The Old English Game Fowl

Old English Game fowl were used for "Cock Fighting" purposes and have been documented in the UK for over 2,000 years since the Roman invasion.

The sport of cockfighting was banned in 1849 after a law was passed which deemed it cruel. Shortly afterwards, poultry shows were beginning to become popular, so although their birds could not legally fight, the breeders of Old English Game were at least able to bench their stock.

There are two main types of Old English Game: the Carlisle type, and the Oxford type, the former being more stocky in build and wider, the latter being of a more upright appearance and with less carriage.

During the cockfighting era, the combs and wattles of males were removed (dubbed) in order to reduce bloodshed and injuries in battle. This tradition continued into the show era, and at present, it is still legal to dub Game and other fowl as long as it is carried out at a certain age.

They say that a good Old English Game is never a bad colour and there are over thirty standardised colours at present. Some of the most popular include: Pyle, Spangled, Partridge, Brown Red, Grey, Ginger, and Wheaten.

Lemon Blue female.

Streaky Breasted Orange Red male.

Spangled female.

Blue 'Brassy Back' male.

A Pyle Old English Game male.

The Pekin

Pekin Bantams are said to have been imported to the UK in the mid nineteenth century. They are very popular in the UK and are a particular favourite among youngsters, no doubt because of their very friendly and tame natures, which make them easy and enjoyable to handle. Being a feather legged breed, they have the tendency to go broody on a fairly regular basis and make excellent mothers.

Pekins are called "Cochin Bantams" in America and other parts of the world and they may appear as miniature Cochins to the independent observer. However those familiar with the two types of fowl will understand that there are significant enough differences for them to be classed as two separate breeds; for example, the Pekin is far lower in carriage than the Cochin fowl. Standardised colours include: Black, Buff, White, Blue, Cuckoo, Lavender, Columbian, and Partridge. However, there seem to be even more non standard colours, such as Silver Wheaten, Red, Mealy Grey, Birchen (Silver and Gold), and Buff-Cuckoo.

A pair of White Pekins.

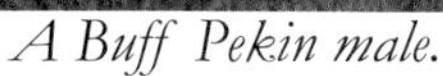

A Buff Pekin male.

A Mottled Pekin female.

A Black Pekin male and a Blue Pekin female.

The Silkie

The Silkie Fowl is generally thought to have originated in China or Japan while others believe, possibly India. It's a unique looking fowl in that it has a woolly appearance, caused by the purity of a recessive gene which prevents the feather barbs from "hooking" together and thus the gene symbol "h" was assigned standing for "hookless". Silkies have five toes and the woolly appearance extends to the shanks, which in the absence of the "h" gene, would be normally feathered.

Silkies have a Blue earlobe, which would normally be white in appearance, as with Hamburghs and other White lobed fowl. However, because the Dark skin genes "Fibromelanosis" are present (Fm), the White pigment is altered and appears as a Bluish cast. White in the earlobe of fowl is caused by the production of "Purine" pigment, and breeds with White earlobes tend to lay White eggs.

Silkies are renowned for their broody qualities and even pass them on to their offspring when crossed to other breeds. They are positively friendly birds, enjoy being handled and make excellent pets for children.

The main standardised colours of Silkie are: White, Black, Blue, Gold, and Partridge. However, other colours exist such as Cuckoo, Red, Lavender, Silver Birchen, and Silver Pencilled.

A White Silkie hen.

A Gold Silkie cock.

A Black Silkie hen.

A Blue Silkie hen.

A Cuckoo Silkie hen.

The Sussex Fowl

The Sussex Fowl is a long established and very old British breed, the original colour being the Speckled. However, by far the most recognisable and popular colour is the Light Sussex with its beautiful contrast of Black and White plumage.

The Light Sussex was thought to have been created primarily by the crossing of Light Brahmas and Silver Grey Dorkings. This would fit with genetic knowledge as the Brahmas would have introduced the genes for plumage and for four toes, and the Dorkings would have introduced the genes for White skin (including shanks), a long body, and single comb. The undesirable genes such as yellow legs and pea combs would have been bred out in time.

The Light Sussex was utilised in the creation of modern hybrids by making use of its "genetically Silver" gene. Thus it was possible to cross Rhode Island Red males to Light Sussex females and achieve "sex identifiable" chick down (see Making use of Sex-Linkage) where the males were Yellow and females were Brown.

The Light gained a great reputation for being a dual purpose egg producing table bird. However, it was separated into two types: the Exhibition Light Sussex, which is far greater in feather, body size and markings, and the Utility Light Sussex, which is far less attractive to look at but has a proven egg laying record.

The Speckled Sussex have enjoyed a recent resurge after some breeders crossed them to Spangled Orpingtons, which released the hybrid vigour. Some strains of Buff Sussex were in decline and consequently breeders made use of the availability to cross to the Light Sussex to revitalise them. This was carried out by crossing a Buff male to Light Sussex females. The result was Sex-Linked offspring, similar to that of the Rhode Island Red to Light Sussex cross.

Other standardised colours include: Silver, White, Red, and Brown Sussex.

Red Sussex male.

Speckled Sussex bantam female.

Light Sussex female.

Buff Sussex males.

CHANGING COLOURS

Black and Red

Exhibition Partridge Wyandotte male.

Cock breeding Wyandotte female.

The Partridge Wyandotte is the closest in colour (of the standardised forms) to that of its natural ancestor, the Red Jungle Fowl. It is therefore logical to see what colours and patterns are attainable by altering either, or both of the residual Black and Red pigments. The original and unmodified Partridge is close to the Exhibition male variety, where the male has a solid Black Breast and females are very mossy in appearance, as pictured above; British and Australian versions have Lemon Hackles, whilst German strains have Orange Hackles.

Without going into the intricacies of what causes the Lemon Hackles, among other differences, it was discovered by Dr W C Carefoot that a single gene - one which he later called the Pattern Gene was responsible for the major difference between Exhibition male and female partridge lines.

The photos opposite illustrate Partridge Wyandottes, which are genetically very similar to the above birds, but which carry, in pure form, copies of the Pattern gene. These are known as Partridge Wyandotte pullet breeders.

Partridge Wyandotte pullet breeding male.

Exhibiton Partridge Wyandotte female.

Pyle and Silver Pencilled

There are only two forms of pigment in poultry plumage and these are Black, and Red. Every variety of colour and pattern seen in poultry today is made by either combining, restricting, intensifying, diluting, or redistributing these two pigments. Without pigment, feathers would be white. With this in mind, I knew I could change the residual colours of the exhibition male or female lines of Partridge and that there were some exciting possibilities.
The photos below illustrate the effects of adding the "Dominant White" genes to a female line of Partridge Wyandottes. Basically, what is happening is that nearly all of the Black Pigment is not being allowed to come to the feather, but the Red Pigment is untouched. This gives the appearance of "Pile" or "Pyle".

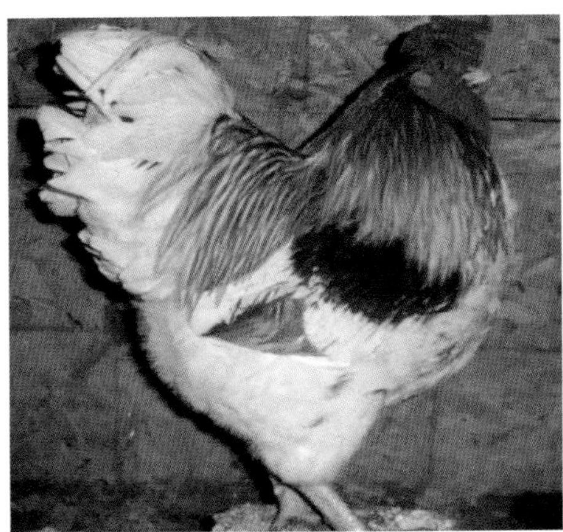

White Partridge Wyandotte male. *White Partridge Wyandotte female.*

As the photos opposite demonstrate, the long established Silver Pencilled variety of Wyandotte is simply just a Silver version of a Partridge Wyandotte. Much in the same way the Dominant White gene restricts Black Pigment from coming to the feather, the Silver gene doesn't allow any Red pigment to come to the feather and hence the Silver Pencilled or as some people call it Silver Partridge effect is possible. The Silver gene is Sex-Linked which I will cover in a later chapter. Sex-Linkage means that males can carry two copies of a gene, whereas females can only carry one. Non Sex-linked genes can be carried in two copies (pure form) by males and females and are called Autosomal.

Silver Pencilled Wyandotte pullet breeding male.

Exhibition Silver Pencilled Wyandotte female.

Blue-Silver Pencilled and Chocolate Partridge

The Blue-Silver Pencilled Wyandotte Bantams pictured below, are examples of a variety that is formed by altering both the Black and Red pigments of the Partridge Wyandotte. The Red pigment is completely inhibited, whereas the Black pigment is only diluted to Blue. Blue-Silver Pencilled Wyandottes exist as Large fowl in the UK, and one noted strain was created by Richard Davies. The Bantams were developed from a single Blue-Silver female imported by Rob Boyd, though an additional line existed and these were created by crossing Silver Pencilled to Blue-Partridge Wyandotte bantams, by Steve Ley and Jim Pearce of Devon.

Blue Silver Pencilled Wyandotte female.

Blue-Silver Pencilled Wyandotte male.

Pictured opposite are the newly recreated Chocolate Partridge Wyandotte Bantams. They were originally created by the late Dr Carefoot, but unfortunately he was unable to continue with them for health reasons and the line died out. The Chocolate gene is just another dilution of Black pigment that seems to have little, if any affect on the residual Red pigment. I created this particular strain over a 4 year period by crossing Chocolate carrying Black Orpingtons with Partridge Wyandottes. Dr Carefoot was aiming for a line of un-patterned Partridge, in relative terms, Exhibition male breeders. The lines featured opposite however, are being bred separately, the male for an Exhibition male line, and the female for an Exhibition female line.

Chocolate Partridge Wyandotte male.

Chocolate Partridge Wyandotte female.

White and Blue Partridge

Adding both the Dominant White and Silver genes to the Partridge, whether it be a male or female line, results in completely White birds. It is important to note though, that this is just one way of arriving at a completely White bird. It is said that there are no fewer than 12 different combinations of genes which will give a White appearance, so in essence, it is impossible to tell just by looking, what a White bird is genetically. It is like covering a painting with whitewash paint. Many breeds are white by the mere presence of two copies of the recessive white gene. This gene basically inhibits all pigment from coming to the feather and the effect is "feathers without pigment", which as discussed earlier, are white in poultry. White Silkies are examples of birds which carry two copies of recessive White.

A trio of large White Wyandottes.

The Blue Partridge, pictured opposite is genetically Partridge Wyandotte or Brahma with the addition of the Blue gene. Basically, the Blue just dilutes the Black pigment. It is proposed that because of the way the Blue gene functions, it is difficult to achieve good pencilling in a pastel Blue shade. Some Blue Partridge females have better pencilling than others, however as the theory suggests, the better pencilled specimens tend to be a darker shade of Blue in the pencilled areas. The Large fowl in the UK were created from Bantams.

A trio of Blue Partridge Wyandotte bantams.

A Blue Partridge Brahma female.

Allan Brooker with a superb Black Wyandotte male.

There is conjecture over the precise genetic make up of yellow legged Black Wyandottes. Many believe they are a Blackened unpatterned Partridge. However, it is my view that Black Wyandottes are genetically very similar to the Silver Sussex with the addition of one or more of the Blackening genes. This opinion has been formed because of the many Birchen and Silver Birchen looking offspring that result when Black Wyandottes have been crossed to other colours. One thing is certain, no matter how many genes are involved in the plumage of the Black Wyandotte, it certainly can be manipulated to create other colours and patterns.

A pair of large Black Wyandottes.

A pair of large Barred Wyandottes.

Adding the Barring gene to Black Wyandottes produces the Barred effect. The Barring gene, like the Silver gene is Sex-Linked which I will cover later in the chapter "Making use of Sex-Linkage". Barred females can only carry a single copy of the Barring gene and hence are always the same colour tone. However, because males can carry two copies of the Barring gene, carriers of a single copy tend to be more like the females, and often display the odd Black feather, whereas males with two copies of the Barring gene are usually a degree lighter, with more White than Black showing in the transverse Bars, making the ratio more like 70 - 30 in favour of White.

An "Impure Barred" male; he only has a single copy of the Barring gene, and hence displays the odd solid Black feather.

The difference in Bar Width between a male with two copies of Barring (left) in comparison to the male on the right which only has a single copy of Barring.

As illustrated above, adding the Blue gene to Barred Wyandottes produces a beautiful pastel Blue-Barred effect. These examples are believed to be the first of their kind and were created by Andrew Bergeson of the UK. The original Blue-Barreds were obtained by crossing a Barred Wyandotte male with a Splash (from Self Blue Parentage) Wyandotte female. The Blue-Barreds lay well, are a good shape and have tremendous hybrid vigour.

Lavender is simply a dilution of Black pigment and is a very attractive pastel and true breeding colour. However, it doesn't come without its problems. For some reason which is as yet unexplained, the Lavender gene has a close linkage to a gene that makes the tail feathers very "raggy" and untidy. The good news is that it's possible to breed out the gene that causes raggy tails, and this has been proven in Pekins, Araucanas, and a few other breeds. Allan Brooker developed the Large Lavender Wyandottes (opposite, below right) by crossing a cross-bred Lavender bird to his Black Wyandottes and making subsequent sibling matings until he was close to his goal. Richard Davies and Margaret Withy developed the Lavender Wyandotte bantams pictured opposite, bottom left. Note the "straw tinge" in the male which would otherwise be Gold in the absence of the Lavender gene; this is something Richard aims to breed out.

Above left a Lavender Wyandotte female expressing purity (2 copies) of the Lavender gene. The birds to the right of her carry only a single dose and hence are Black.

Cream

As the photos below illustrate, adding the Cream gene to a Partridge Wyandotte produces the Gold Duckwing effect, for want of a better name. However, Gold Duckwing Wyandotte males can be easily confused with Impure Silver Pencilled Wyandotte males. If you study the photo below (left) of the Gold Duckwing male, you will note that the Impure Silver Pencilled male (oposite) is very similar in appearance, but what each male will breed is quite different.

Knowledge is power with poultry, so it's important to keep records of which stock came from each pen. The Impure Silver Pencilled male is produced by crossing Partridge to Silver Pencilled (can be either way round). Not all resemble the Impure pictured opposite, some are quite devoid of Red but usually display a "wilted" White rather than Clean Silver.

Breeding to Partridge Wyandottes would reveal the true genetic identity of each respective male. The Gold Duckwing bred to Partridge would produce only Partridge offspring, all carriers for Cream. However, the Impure Silver male would produce: 25% Partridge Pullets, 25% Silver Pencilled Pullets, 25% Partridge Males and 25% Impure Silver Males (as is he). The Cream gene is found in many Silver varieties of fowl and even emerges in lines of Large Partridge and Buff Columbian Wyandottes.

A pair of Gold Duckwing Wyandottes
(a better name could be 'Cream Partridge').

An impure Silver Pencilled Wyandotte male.

A 'Cream Columbian' Wyandotte female,
a sport from a Buff Columbian pen.

Adding the Columbian gene to an Exhibition male line of Partridge Wyandottes (pics A & B) produces the "Buff-Columbian" effect (pics C & D). However, it is not to be assumed that adding Columbian to an

Exhibition female line of Wyandottes (pic J) would produce birds with the same pronounced neck markings. It is my view, and that of many genetics enthusiasts, that a gene with the proposed name "Hackle Black" is required to be present in Partridge Exhibition male lines, and that the make up of all Columbian varieties requires more than the presence of Columbian, and the lack of the Pattern gene on each respective background (see the chapter Hackle Black Theory).

It was thought for many years that simply adding the Pattern gene to an Exhibition male line of Partridge would yield Partridge females closer in resemblance to the Exhibition females. However, in my view, adding the Pattern gene to an Exhibition male Partridge line would produce birds close in appearance to the Gold (pic G) and Dark Brahma (pic H) females. It would therefore seem logical that there does exist a Hackle Black gene and its presence is essential in Buff Columbian and Exhibition male Partridge lines, and even in Light and Buff Sussex fowl. However, just as important is its absence in the Exhibition Partridge female lines (pic J). I believe that pic E, demonstrates fowl which either carry or are devoid of the unofficially proposed Hackle Black genes.

Hackle Black.

Non Hackle Black.

Hackle Black.

Non Hackle Black.

The Light Brahma and the Buff Columbian Wyandottes (pics K & I) show clearly the presence of the proposed Hackle Black and how it plays its part in the make up of both varieties. The Partridge and Silver Pencilled Wyandottes (pics J & L) show clearly the absence of Hackle Black and how its absence allows for a neck that is more finely pencilled than that of the Dark or the Gold Brahmas (pics G & H).

Large Light Sussex females.

Columbian Wyandottes

Above: Columbian males.

Right: Columbian Wyandotte pullets which also rely on the presence of the Hackle Black genes for their striking neck markings.

Large Columbian Wyandottes were re-made by some breeders by crossing Light Sussex to Silver Pencilled Wyandottes and selecting accordingly for the desirable features from sibling and subsequent matings.
The only major difference between the plumage genotype (genetic make up) of the Columbian Wyandotte and the Light Sussex is their genetic foundation. Columbian Wyandottes are genetically Brown "eb" whilst Light Sussex are Wheaten "eWh". This accounts for the differences between the two, such as the Columbians having a Dark under-colour and the Sussex having a light under-colour. The presence of the Brown allele (eb) in the Columbian Wyandottes accounts for the saddle striping in the males. Light Sussex males are generally devoid of saddle striping, but some birds are Wheaten/Brown (eWh/eb) so some striping occurs. In such birds, the under-colour is generally White from the base, changing to grey half way up the feather.
Other than the aforementioned differences, Columbian Wyandottes and Light Sussex share the same genes for plumage and hence are visually very similar; these genes are Columbian, Silver, and the proposed Hackle Black.

Creating the White-Buff Columbian Wyandottes (or what I named "Vanilla") was relatively easy. It was simply a case of crossing a Buff Columbian male with some Pyle Wyandotte females. All I had to do then was cross the prototype first generation Vanillas back to Buff Columbian Wyandottes, each time selecting for the 50% Vanilla offspring. This process improved type but also "purified" the Columbian genes within the Vanillas. If you look at the photo of the three growers (above, lower right) you will see that the female on the far right is pure for Columbian genes (she has both available copies), whereas the female on the left only carries one of the available Columbian genes and hence is more white on her back - the White would show as Black "moss" if the pullet was devoid of the dominant White genes. A Vanilla Wyandotte is basically a Columbian Wyandotte with Dominant White.

The original Vanilla male being bred back to Buff Columbian females.

Chocolate

A Chocolate Orpington pair.

The Sex-Linked recessive gene named "Chocolate" by the late Dr Carefoot in 1995 is another dilution of Black pigment. The Chocolate gene is one of the few known Sex-Linked recessive genes in poultry. It may seem difficult to grasp at first, but if a female carries Chocolate, then she will be visually Chocolate. However, males require two copies of the gene for them to appear as Chocolate. Black males can carry a copy of Chocolate without it being at all identifiable. The only way to tell is to test mate. However if the mother of the Black male in question was Chocolate herself, it is a certainty that all her sons will carry a single dose of the Chocolate gene.

Chocolate to Chocolate gives: 100% Chocolate offspring
A Chocolate male to a Black female gives: 50% Chocolate females. 50% Black males (the males will all carry Chocolate)
A Black male to Chocolate females gives: 50% Black males (Chocolate carriers) 50% Black females (normal Black)
A Chocolate carrying male to a Black female gives 25% Chocolate females, 25% Black females, 25% Chocolate carrying males, 25% Black males - though it will be impossible to tell which males carry chocolate.
A Chocolate carrying male bred to a Chocolate female gives: 25% Black females, 25% Chocolate females, 25% Chocolate males, 25% Chocolate carrying males (Black).

A Black female to the right for comparison.

A Black male to the left for comparison.

A Chocolate Wyandotte female (prototype).

A Chocolate Orpington male (trimmed for breeding).

An all Chocolate breeding pen.

Lavender-Columbian / Coronation

After some help with stock from Margaret Withy, and some advice sought from me, Richard Davies decided he wanted a true breeding "Coronation Wyandotte" and so was fortunate enough to breed a pullet that resembled in many ways what he had envisaged (although she was very mossy on the back - see opposite pics). For the Coronation, or perhaps in this case a better term Lavender-Columbians, to breed true, Richard knew it would require the purity of both the Lavender and Columbian genes.

Unfortunately, Richard experienced an unexpected fox attack that happened during the daytime and he lost his Lavender-Columbian looking female. However, he was lucky that he'd already hatched some twenty plus chicks from her, the result of her pairing with a Columbian Wyandotte male. This mating worked well, as it helped to produce specimens with good, clear White back feathers which resembled good Columbian Wyandottes.

Because Lavender is a recessive gene, Richard knew that he would have to carry out a further sibling mating to extract the Lavender genes at a rate of 25%. However, once these Lavender-Columbians were obtained, they would breed true, unlike many Coronation Sussex strains which rely on the Blue gene for their pastel shade and hence produce 50% Coronation, 25% White, and 25% Light Sussex offspring. Only Wyandottes were used in this project.

Comparing hackles (a Lavender Columbian on the right).

A Coronation Sussex male.

The original Lavender-Columbian female.

The Lavender-Columbian hen with a Columbian male.

The Wyandottes pictured above, which I call Blue Salmon, were made by crossing Wheaten Marans to Partridge Wyandottes, then breeding an Impure Silver Pencilled male to the female Wheaten offspring (F1), and selecting an Impure for Silver and Wheaten male (F2), then crossing him to a rare large Blue-Silver Pencilled Wyandotte hen (see bottom right). The initial aim was to bring the Wheaten genes into a line of Large Wyandottes and breed from a pen that would segregate the Silver and Blue genes, meaning it would be possible to produce Blue Salmon, Wheaten, Silver Wheaten, and Blue

Prototype Wheaten Wyandottes.

The F2 male crossed to a Blue-Silver Pencilled female.

Wheaten from the same pen. However, I was so taken by the Blue Salmon that I decided to pursue them separately and enlisted the help of friends to carry them on as well. Other friends carried on my lines of Wheatens, and Silver Wheatens with a great deal of success.

It is usually very difficult to achieve Red shoulder feathering on genetically Silver males, that is unless the autosomal gene 'Ap' is present, which it clearly is in Wheaten varieties such as the Salmon Faverolle. Salmon Faverolles are basically Silver Mahogany Wheatens. I knew it would be possible to create Blue Salmon by bringing in the Wheaten genes (from the Marans), and that it would be a very attractive colour. It was easy to select for the Wheaten carrying chicks because they were predominantly yellow with faint striping, as opposed to the clear yellow down of pure Wheaten chicks.

A Blue Salmon male.

Bredding pen A. *Progeny from breeding pen A (aside from non Buff birds).*

Male and female in breeding pen B. *A son from breeding pen B.*

A son from breeding pen B and a daughter from breeding pen A are crossed together.

Buff Wyandotte male and females resulting from the pair at the bottom of page 70.

Self Buff birds are based on Wheaten (eWh), because it is the basis or "e allele" (see the chapter, "e+ In the Beginning") which naturally has the least amount of Black. Buff has a host of Red pigment extending genes which all have to be pure to make a solid Buff fowl. There is debate over the precise number of Red extending genes and what they are, but the common consensus is that Self Buff birds require Wheaten (eWh), Columbian (Co), Dark Brown (Db), Dilute (Di), and Champagne Blonde (Cb). The genes which allow for a completely Buff tail, are as yet to be fully determined. There are obviously different shades of Buff, so there isn't necessarily a correct, or incorrect formula.
However, one thing is certain: breeding a self Buff bird to a bird which doesn't carry any or all of the genes required for solid Buff, will in most cases result in Black tailed Buffs.
From the Buff-Columbian matings, an unexpected female arrived. She looked Buff-Columbian but lacked the necessary Hackle Black markings and so I thought she could be utilised to recreate a strain of Large Buff Wyandottes. A Large Buff Rock was obtained to be mated with her, which yielded 100% Black tailed Buffs, as expected. These birds all displayed Rose combs (MatingA). A separate mating of a poor Buff Wyandotte to Buff Orpingtons yielded 100% White legged clean Buffs, as expected (MatingB). In order to achieve a clean Self Buff Wyandotte, I could have attempted a sibling mating from the F1 from MatingA. However, I felt that this would have yielded too many Black tailed Buffs and consequently would have produced too much waste. It seemed plausible that if one half of the breeding pen was pure Buff and the other half had at least half the genes for Buff, then at least 50% of the offspring would emerge as pure Buff. This certainly was the case and a breeding pen made of a male from MatingB to a female from MatingA, produced 50% Pure Buffs. Obviously, White legs and Single combs were produced, however enough clean, Yellow Legged, Rose Combed Buffs were produced which allowed for an all Wyandotte breeding pen.
This pen, as expected, produced true to type, with 100% Yellow Legs, and Rose Combs (though some may carry the single comb gene). There was a little variation in shade of the Buffs produced, however as per usual, selection was the key and some very satisfactory results were reported. The strain is still producing well and is being shown with a lot of success in the UK.

Buff is the most difficult colour to introduce to a new breed, simply because it has so many genes involved to make the full effect. It was possible with the Buff Wyandottes because there already existed a strain (albeit poor) of Buff Wyandottes and type didn't require too much selection as Wyandottes were used in the initial crosses, which were then blended together.

It would be very difficult (but not impossible) to transfer the Buff colour onto a breed that doesn't currently have Buff as a colour option. Because all the genes for the full Buff effect require purity, it is more than possible to breed a Buff bird to another breed and then carry out a sibling mating and produce a degree of clean Buff fowl. However, the chances of carrying out that process and producing birds with the type of the desired breed straight away are very slim indeed.

It is a far more realistic prospect to carry out the above process and then use the closest in type fowl, F2 (which is clean Buff) back to the desired breed; the process would have to be repeated, and perhaps carried out again, with a possible total of 6 matings before anything began to actually resemble the desired breed in the Buff colour.

Buff Wyandottes - the result of crossing the progeny of matings A and B.

The Complexity of the Blue gene

The Blue gene (Bl) varies greatly in expression, from a pastel "powder puff" Blue to a mid-Blue, right through to a very charcoal Blue, with many shades in between these three examples.

The Blue effect is achieved by a "widening" of the Black pigment granules present in the feather structure, and is seen through an opaque layer of protein keratin.

In such varieties as the Andalusian, the Blue Orpington, and the Blue Australorp, a distinct dark lace is required to surround the blue interior of the feather. This effect was thought to be made possible by the presence of the combined Lacing genes of the Laced Wyandotte being present in an otherwise self Blue fowl (Carefoot 1984).

The Blue gene was always thought of as largely "unpredictable" in relation to the particular shade it produced when present in single form. The Mendellian laws of the Blue gene have long been established and two copies of Blue gives a splashed appearance, no matter on what background.

So why the varying shades? In my opinion, the Blue gene is consistent in expression but relies heavily on the quantity of Black pigment present in any given fowl to determine its particular shade. If such Black genes are purified within a strain it is likely that the shade of Blue produced will be consistent, regardless of the particular shade itself.

Black male x Splash female
=
100% Blue offspring

A LOOK AT THE GENES

In the Beginning, the e+ (Wild Type)

It is clear to see from the photos opposite that Breeds such as the Dutch Bantam and the Brown Leghorn derived from the Red Jungle fowl as did all fowl which is the generally accepted assumption. Whether this thinking is correct, that all progressive genes and traits seen today in all fowl mutated one by one from the Red Jungle Fowl, or whether the new thinking is correct that hybridisation of one or more of the four types of Jungle Fowl is a more convincing theory, remains to be proven. However, one thing is for sure, the Red Jungle Fowl had an enormous part to play in the development of modern fowl and allowed for the application of Mendel's Law of Genetics in poultry breeding, in order that we can understand how the genes pertaining to most traits function.

There are only two forms of pigment in poultry, Black, and Red. Red describes a host of warm tones from the palest Lemon right through to the deepest Mahogany. If you study the birds opposite, you will see that they are made up of Black and Red Pigments; the Brown parts (on the backs of females) are caused by a mixture of the two pigments

The Wild Type colouring is nature's way of distributing Black and Red pigment, no doubt to make the males attractive to females and to make females a lot more subtle in plumage for the purposes of camouflage when incubating their eggs.

The "e" allele sounds complex, but it is just a scientific way of referring to how the Black and Red pigments are distributed in a fowl. There are 4 other generally accepted alternatives to the Wild Type colouring (colour bases if you like). The good news is that the Laws of Mendel can be applied with these alternatives and they work in exactly the same way as Dominant and Recessive genes (see the chapter, Dominant / Incomplete Dominant / and Recessive genes), so crosses between birds with different "e" alleles can be easily calculated with knowledge of their Dominance and the use of Punnett Squares (see the above mentioned chapter)

Before a fowl can exist as any colour or pattern, its plumage (primary) has to be regulated by one of three options: the original "e" allele (Wild Type), one of the four accepted alternatives (Black, Gold Birchen, Wheaten or Brown), or a

Red Jungle Fowl male.

Gold Dutch male.

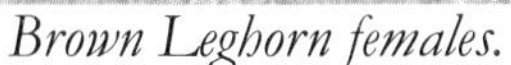

Brown Leghorn females.

Red Jungle Fowl female.

combination of two of the five possibilities. Just as with all Autosomal (Non Sex-Linked) genes, each parent contributes one allele to each of its offspring. We've discussed how the Wild Type allele distributes Black and Red pigment, but what about the four others?

The standard alternatives are: Extended Black (E), Gold Birchen (ER), Wheaten (eWh), and Brown (eb).
There is an order of Dominance, when these alleles are bred together, and the order of Dominance is:
E > ER > eWh > e+ > eb
E stands for Extended Black and most Black fowl are E at the "e" locus, which extends pigment into the shanks. However, most Black fowl require the presence of at least an additional Black gene such as Melanotic to make them fully Black. Examples of birds which are E/E are Black Langshans, Black Orpingtons and Black Sumatras.
ER stands for Extension with Restriction, so it acts in much the same way as E, but allows for some Red pigment to come through in the shoulder, hackle and saddles of males and in the hackles of females, with varying lacing in both sexes (what we know as Gold Birchen). Silver Birchen varieties just have the Silver gene added. Such an example is the Silver Sussex. An Example of Gold Birchen (ER/ER) is the Brown-Red Game. There are currently thought to be two types of ER, one which allows for dark shanks and the other which allows for clear shanks (free of Black pigment).
eWh stands for extension of Wheaten. Males are very similar to the Wild Type but have a distinct White undercolour. Females are very Salmon in colour and have a "clay" tone with a predominantly Red pigment spread throughout the whole body, especially on the back. Examples of eWh/eWh birds are: Wheaten Game, Shamo Game, Wheaten Maran, Buff Fowl (underneath the modifying genes such as Columbian etc).
e+ is the unmutated "e" allele (Wild Type). There are thought to be two separate types, very close in appearance, but which interact the same with the other "e" alleles.
eb stands for Extension of Brown. Males are very similar to the Wild Type but females are devoid of any Salmon colouring in the breast, which makes for an all over stippled and mossy fowl. Adding the Pattern gene to the eb allele produces the concentric pencilled Partridge. Examples of birds which are eb/eb, without modifiers such as the Pattern gene (Pg) and Columbian (Co) are: Partridge Wyandottes, Exhibition Male Breeders.
Without any plumage modifying genes such as the Pattern gene (Pg) or Columbian (Co) to skew results, a hybrid fowl has to be a combination of two of the above "e" alleles (one from each parent) and will look mostly like the

Top: An F1 male from a Welsummer male x Barred Wyandotte female.
Middle: An F1 female from a Welsummer male x Partridge Wyandotte female.

The result of crossing Partridge Exhibition lines together.

Dominant parent, which is indicated depending on which of the two alleles is nearer to the left of the order ofdominance sequence- E>ER>eWh>e+>eb- (most Dominant). For example, a bird that is eWh/eb (a cross between a Wheaten Maran and Partridge Wyandotte), will be predominantly the colour of a Wheaten Maran. However knowledge serves as a reminder that the bird in question is genetically half Partridge.

Crossing two pure breeds together is one way of discovering which "e" allele is most dominant. Obviously secondary genes which affect plumage (such as Barring, Pattern Gene, Columbian etc) will have an influence on the colour of the offspring, as outlined above.

I located a Welsummer male, and first crossed him to a Barred Wyandotte female.

The resultant male's genetic make was E/e+ but with a single copy of the Barring gene (see top picture on page 79). We know he is not E/E as his mother was, because his father was a Welsummer (e+/e+), thus making him E/e+.

The same Welsummer male was used on a Partridge Wyandotte female. The resultant female's genetic make up was e+/eb (see top picture on page 79).

So it's easy to see how the different "e" alleles interact with each other. Both fowls have just a single copy of a secondary plumage gene: the male has a single copy of the Barring gene, and the female has a single copy of the Pattern gene, introduced by her mother. However, it is easy to observe that the Pattern gene (Pg) cannot really express on the e+/eb primary plumage, especially in a single dose; it requires the presence of both copies of the gene and on a primary background "e" allele of eb/eb, as are Partridge.

The pullet pictured (see bottom picture on page 79) is a direct result of crossing Exhibition male and female lines of Partridge Wyandottes together. While she is pure for her primary plumage - the "e" allele, eb - her mother was pure for the secondary plumage gene, the Pattern gene (Pg), whilst her father was pure for the absence of the gene (pg+). This makes the offspring in question eb/eb Pg/pg+. It's not wise to make such a cross because as you can see, the pattern produced resembles neither parent and is neither one thing nor the other.

Building a Variety

The sequence of photos below from A to F shows how a Blue-Laced Silver Wyandotte is genetically "eb" (primary plumage). First the Silver gene is added (pic B), followed by the Pattern gene (pic C), then the Melanotic gene - Black (pic D), the Columbian gene (pic E), and finally the Blue gene (pic F).

Creating a New Breed or Colour

Is it wise to create a new breed or "colour variety" within a breed of poultry? No doubt opinions will differ, and I understand the point of view of those who don't agree with the crossing of different breeds or colours of poultry. We do have plenty of choices when it comes to colours, shapes, features and varieties of feather plumage within the available poultry breeds. However, many people believe that it's perfectly acceptable to create a new colour or breed, using the example of "evolution" as their main source of reasoning, and that everyone can "do as they so wish".

Whilst this may be true to a point, we must remember that without the breeders who have between them kept the strains of exhibition poultry pure for centuries, we would undoubtedly have no way of quantifying and understanding the genes in the way we do today. Breeders are owed a lot of credit for their efforts and perseverance. So with this in mind, the desire to create something new, especially if you intend to introduce it to the show world, should be given long and careful consideration. Crossing different varieties or colours of poultry together in the back garden, just for fun, is quite a different matter.

Ideally, the creator of a new breed, or colour should focus solely on the project he or she is about to embark on and not be distracted, or restricted by the amount of space available by keeping other breeds. Creating something new should not be entered into lightly. More often than not, it takes a lot of time, money, space, effort, patience, resilience, and determination.

I have created many new colours of Wyandottes over the years, but I entered each project with a realistic understanding of just exactly what it would take to arrive at each goal. Many projects took at least four years to produce something that resembled my vision, which in truth was just the starting point of each new variety.

The reason I mentioned the requirement of "space" earlier was because new varieties often necessitate the selection of a small percentage of desirable offspring in relation to the amount of offspring produced in each given generation. Hence usually a large number of chicks need to be hatched to give the breeder more chance of producing the small percentage of desirable chicks, which are necessary for each progressive step of the project.

To use an example, the Chocolate-Partridge project was ambitious from the offset. My intentions were quite complex. Because I wanted to breed two lines (the respective Exhibition male and female lines) - as with normal Partridge, I began by crossing a Chocolate Carrying Black Orpington male to Exhibition Partridge females. However, I also wanted to breed a separate line of what I call Coffee-Creams (basically Chocolate Partridge with the Cream gene added).

The first cross (F1), as expected yielded:

50% White Legged Rose Comb Black males - of which half carried chocolate (discarded)

25% Black Legged Rose Comb Black females (discarded)

25% Chocolate Legged Rose Comb Chocolate females (kept for step 2)

As I had prepared for, only 25% of the chicks hatched were desirable, which was a difficult factor to accept, but commonplace in such projects.

The next step was to breed a Gold Duckwing Partridge male to the F1 Chocolate female (to bring in the Cream genes).

This produced the F2, which consisted of:

12.5% Gold Birchen looking males (discarded)

12.5% Black Males (discarded)

25% Partridge males (Kept for step 3)

25% Partridge females (discarded)

12.5% Black females (discarded)

12.5% Gold Birchen looking females (discarded)

In step 3, I bred an F2 Partridge male which I knew had to carry the Chcolate gene (because his mother was Chocolate) to respective Exhibition cock, and pullet breeding Partridge females. The results were as follows:

25% Chocolate-Partridge females (Kept)

25% Partridge females (discarded)

25% Partridge males (discarded)

25% Chocolate carrying Partridge males, indistinguishable from Partridge males (discarded)

This is where the project got interesting. Instead of becoming easier, it became more complex. At this point, I knew 50% of the Chocolate-Partridge (F3) females would be carrying the Cream gene and that bred back to their father, would produce 50% Gold Duckwing Offspring - whether visually Chocolate or not. Before I could carry out this mating, which would more importantly produce some Chocolate-Partridge males, I had to sort through a great many factors which I hadn't accounted for. These included single combs, and a variety of dark leg colours (not just yellow). The dark legs were attributed to a Sex-Linked recessive gene carried by the F2 Partridge male (id+).

The next step was to cross the F2 Partridge male to his F3 Chocolate-Partridge daughters.

This mating yielded some satisfactory results, and even produced some Coffee-Creams as well as Partridge and Chocolate-Partridge males and females. However, many birds displayed additional unwanted factors which included varying shades of dark leg colour, and / or single combs. Selection was indeed the key in this mating and the approximate results were as follows:

7% Partridge males - carrying Chocolate (discarded)
7% Partridge females (discarded)
7% Chocolate-Partridge males (kept)
7% Chocolate-Partridge females (kept)
7% Coffee-Cream males (kept)
7% Coffee-Cream females (kept)
7% Black males (discarded)
7% Black females (discarded)
7% Chocolate males (discarded)
7% Chocolate females (discarded)
7% Birchen looking males - to include Cream versions (discarded)
7% Birchen looking females - to include Cream versions (discarded)
7% Chocolate-Birchen looking males - to include Cream versions (discarded)
7% Chocolate-Birchen looking females -to include Cream versions (discarded)

From the above statistics, it is clear to see that the F2 Partridge male inherited some fairly distinctive Blackening genes from his F1 Chocolate mother, and that although only the Sex-Linked recessives became apparent in the F3 offspring, it was possible to reveal the true genetic identity of the F2 Partridge male by breeding him to his daughters. Considering the fact that such a small percentage of desirable stock was produced from the F4, things were further hindered by the emergence of birds with dark legs and single combs. These factors severely restricted selection and meant that the original recreated Chocolate-Partridge male unfortunately displayed Chocolate legs. However, these were bred out in the F5 generation.

Chocolate carrying Orpinton male x Partridge female. Pen A

Chicks from pen A, Chocolate chicks left and Black chick right.

B -An F1 Chocolate female Daughter of pen A.

C - Gold Duckwing male.

D - Partridge male, Chocolate and Cream Carrier, result of pen B x C.

E - Chocolate-Partridge male, the result of male D x female F.

F - Chocolate-Partridge female, the result of male D over a Partridge Wyandotte female.

The many chicks produced.

Dominant / Incomplete Dominant / and Recessive Genes

You will likely have heard the expressions "Dominant" and "Recessive' genes and you may even have some experience with them in poultry, but what are they, and why do they have to operate in this manner? "It sounds very complex" is what I often hear.

Many traits in poultry are caused by the action of a particular gene, present in pure form (2 copies). If you learn that the maximum amount of genes that can be carried for each respective gene (not trait) is 2, then you are onto a good start. For example, the maximum amount of Lavender genes a bird can possibly carry is 2, by inheriting one copy from each parent. The least amount of Lavender genes a bird can carry is 0 by inheriting no copies from its parents.

The reason I say the maximum amount of genes that can be carried for each gene and not trait, is 2, is because traits such as feathered legs on Brahmas for example require the presence of at least 3 separate leg feathering genes - all present in pure form to have full effect.

The term "Locus" (Loci plural) sounds complicated, but all it represents is the place on a fowl (chromosome) where a gene is either carried or not. Each fowl has 2 Loci for each gene (so 2 available locations, should it encounter the gene). This applies across the board with the exception of Sex-linked genes which are carried on the sex chromosomes.

Each gene will express fully if it is present in 2 copies and is not suppressed by another gene. Why should this happen? It's basically because some genes are "stronger" than other genes and are termed "Dominant". The weaker genes are known as "Recessives". Let's cross a Partridge Wyandotte to a Partridge Cochin, and let's assume we're working with pure strains. The resulting offspring (F1) would all be feather legged, rose combed birds.

You may wonder why it works this way. Basically the Wyandotte has 2 locations (Loci), one from each parent, which carry the message for clean legs to develop. By crossing to a Cochin, the offspring still have one Loci with the same message as the Wyandotte, but one with the same Loci as the Cochin which carries the message to let feathers develop on the legs. As has been sequenced, the leg feathering genes in Brahmas and Cochins are the cumulative effect of genes on three separate Chromosomes. Since feather legged offspring are always obtained from crosses with Brahmas and

Cochins to clean legged fowl, the leg feathering genes have been assigned as Dominant, and the clean legged genes, Recessive.
However, it has been noted that subtle leg feathering genes are regarded as Recessive as well. This is because breeders of clean legged stock have been known to experience the odd feather legged fowl emerging from their lines, but they are usually very sparsely feathered.
The Wyandotte-Cochin crosses, as well as having feathered legs, would also display rose combs. This is because the rose comb gene is stronger than the single comb gene in expression. However, it's not sufficient to say that rose combs are dominant to single combs and leave it at that. It would be quite easy for an inexperienced person to cross a bird such as the Wyandotte-Cochin F1 to a line of single comb birds, experience 50% rose combs and 50% single combs in the offspring, and then be quite sceptical about the whole notion of rose combs being dominant to single combs.
To make it easy to calculate, the geneticist, Reginald. C. Punnett came up with the idea of using squares to predict the outcomes of mating together birds that weren't pure for the same genes. The more genes involved, the more complex the squares become, and they only work in relation to "Mendel's Law of Gene Segregation". Not all genes work this way, however most of them do and understanding this principle goes a long way to understanding genetics in general.

A Partidge Wyandotte male (clean legged).

Gold Brahmas (feather legged).

Correct Genetic terminology is detailed in the back of the book - to avoid complex looking symbols, I will demonstrate using the following:
A pure rose comb bird has 2 copies of the rose comb gene and is R/R
A pure single comb bird has 2 copies of the single comb gene and is r/r
If we cross them together, all of the offspring become visually rose combed, but are impure for the rose comb genes. If we crossed the siblings (F1) together, we would have gametes of R/r x R/r (each bird carrying equal genes - 1 for rose, and 1 for a single comb)

	R	r
R	R/R Pure Rose Comb 25%	R/r Visually Rose comb but Impure like each parent 25%
r	R/r Visually Rose comb but Impure like each parent 25%	r/r Pure Single Comb 25%

By writing the Gametes out in the way shown above, it is possible to predict the outcomes of such a mating. Simply put in each box the letter you see in the top and margin Bars (keeping the capital letters to the left to indicate dominance), and it's easy to see that crossing two birds together which are impure for rose combs (such as the Wyandotte-Cochin F1), yields 75% visually rose combed offspring
The Gametes in relation to comb of a Wyandotte-Cochin bred back to pure Wyandotte would be R/r R/R
The Gametes in relation to comb of a Wyandotte-Cochin bred back to pure Cochin would be R/r r/r
Just by entering these Gametes into the Punnett square, it is possible to predict the outcomes of each respective mating.

The circles below show a pair of Loci, in their original state (left), with a single copy of the Blue gene (middle), and with two copies of the Blue gene (right). Below each combination is the colour of bird that it produces. It is easy to see why a bird with either of the left or right Loci (Black or Splash), paired to a partner carrying the same genes, would breed true. However, the Blue gene in pure form, produces predominantly White colouring in feathers where Black pigment would usually be. The effect we call Blue is basically a gene in it's "Impure" state and hence that's why it doesn't breed true. It is classed as an "Incomplete Dominant" gene because it's effects can be seen in single form, but quite clearly, not the same as when in its pure form, which produces the "Splash" effect.

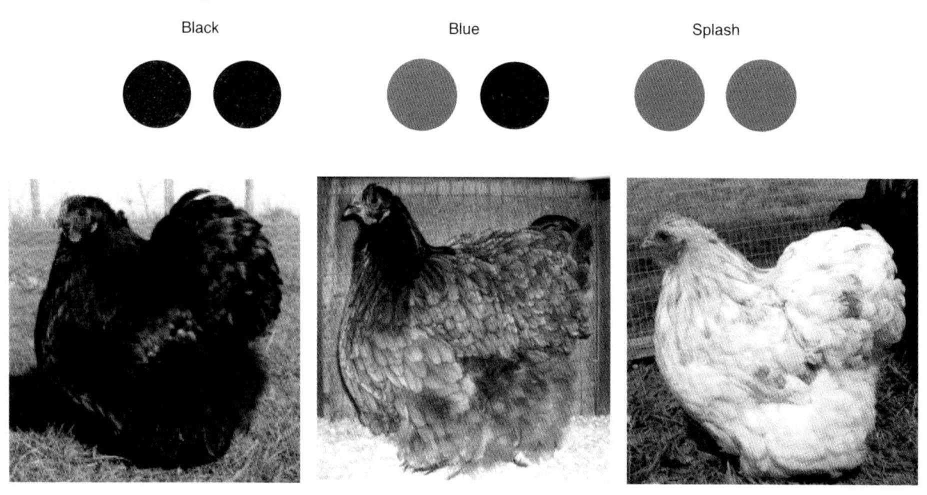

Blue Male

	Black Birds 25%	Blue Birds 25%
	Blue Birds 25%	Splash Birds 25%

Left is a Blue Female

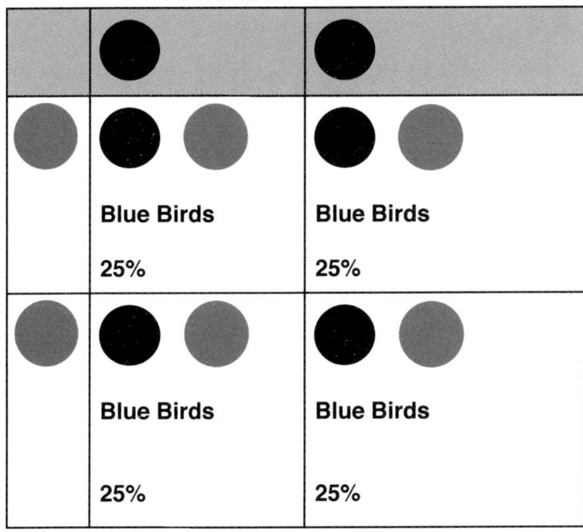

Left is a Splash Male or Female

Black Male or Female

	Black Birds 25%	Black Birds 25%
	Blue Birds 25%	Blue Birds 25%

Left is a Blue Male or Female

As the above Punnett squares demonstrate, by writing in the "Gametes" (or Loci) for a trait from each respective parent, in this case the Blue gene - it is possible to accurately predict the percentage outcomes of the offspring produced in each given combination.

Perhaps another and possibly easier way of explaining how the Blue gene works is to use the example of "gloves". If we use people and gloves as examples of genes it will become clear. Imagine that the person who represents a Black fowl is wearing a pair of Black gloves and that a person wearing a pair of Blue gloves represents a Splash bird. Now imagine that each individual was asked to place one glove on the table - from either hand. There is only 1 possible outcome and that is a Blue and Black glove would be placed on the table - 1 from each person. This outcome is easy to predict as it can only go one way and that is why crossing Black to Splash birds always results in 100% Blue offspring.
Now, imagine a person with 1 Black glove and 1 Blue glove is put in a room with a person wearing exactly the same as them (both represent a Blue fowl). As with the previous example, both people are asked to place a single glove on the table. Can we predict the result? Well it's not as clear cut as the first example. This time, and with each time they do so (each offspring produced) there is a possibility of 1 of 3 combinations on the table, there could be: 2 Black gloves, or 1 Black and 1 Blue glove, or 2 Blue gloves. But what are the chances of those 3 possible outcomes? Well considering both parents have a Black and Blue glove each, the chances are that 50% of the time, there would be a Black and a Blue glove on the table (Blue fowl), there's a 25% chance of there being 2 Black gloves on the table (Black fowl) and a 25% chance of there being 2 Blue gloves on the table (Splash fowl).
I hope the above hypothetical examples will give you a better understanding of how an Incomplete Dominant such as the Blue gene works.

A Lavender Wyandotte pullet (simular to Blue but breeds differently).

A Blue Wyandotte.

What goes in must come out...Surely?

In many cases where a new breed or colour form is being created, it is necessary to make the initial cross which will produce the F1. It is usually a case of then carrying out a sibling mating, so crossing the F1 brothers and sisters together in order to release the Recessive genes in the next generation (the F2). An example here could be the Lavender gene. Because it's Recessive, (just like the Single comb gene), it would be necessary to cross a Black fowl to a Lavender fowl and then carry out a sibling mating in order to produce 25% Lavender fowl in the F2.

No doubt the reason for doing this would be to transfer the Lavender genes from a particular fowl onto something more desirable. An example is when Allan Brooker bred a Lavender Orpington cross to his Black Wyandottes. He then had to cross the F1 together in order to release the Lavender genes but with birds that displayed rose combs. This process was repeated with subsequent generations of Lavenders being bred to Blacks, and then sibling mated until type and leg colour resembled a Lavender Wyandotte.

Breeding in Recessive genes and then selecting for them at a rate of 25% when the F1 are crossed together is one thing. However, as in Allan's case, when selecting for two separate traits (rose comb and lavender) the percentages are clearly reduced.

Opposite is a Punnett square which calculates the recombination percentages of 2 traits. Note how much more complex it is than the ones on page 90 which just calculated the recombination outcomes of a single trait.

Minorca Single Combed.

Wyandotte Rose Combed.

An F1 from Lavender Orpington x Black Wyandotte

	R Black	R lav	r Black	r lav
R Black	R Black 1 R Black Pure Rose combed Black Birds 6.25%	R Black 2 R lav Pure Rose combed Blacks carrying Lavender 6.25%	R Black 3 r Black Pure Blacks with Rose comb carrying single combs 6.25%	R Black 4 r lav Rose combed Blacks carrying single comb and Lavender 6.25%
R lav	R lav 5 R Black Rose combed Blacks carrying Lavender 6.25%	R lav 6 R lav Pure Rose combed Lavenders 6.25%	R lav 7 r Black Rose combed Blacks carrying single comb and Lavender 6.25%	R lav 8 r lav Pure Lavenders with Rose comb carrying single combs 6.25%
r Black	r Black 9 R Black Pure Blacks with Rose comb carrying single combs 6.25%	r Black 10 R lav Rose combed Blacks carrying single comb and Lavender 6.25%	r Black 11 r Black Pure Single combed Blacks 6.25%	r Black 12 r lav Pure single combed Blacks carrying Lavender 6.25%
r lav	r lav 13 R Black Rose combed Blacks carrying single comb and Lavender 6.25%	r lav 14 R lav Pure Lavenders with Rose comb carrying single combs 6.25%	r lav 15 r Black Pure single combed Blacks carrying Lavender 6.25%	r lav 16 r lav Pure single combed Lavender Birds 6.25%

In the far left column is an F1 from Lavender Orpington x Black Wyandotte

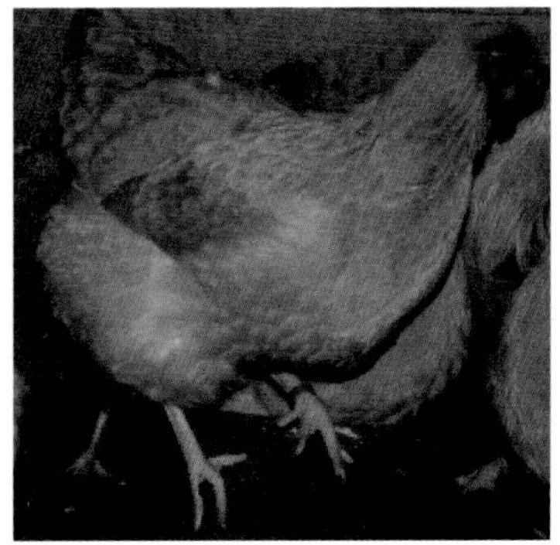

In reality, when crossing two fowl together, each carrying pure genes for separate traits, you are splitting the genes in half in the F1. Sibling matings are carried out specifically to recombine these genes in the correct and desired order. As with the Lavender Orpington to Black Wyandotte cross, a mating of the F1 offspring only produced 6.25% Lavender birds which were pure for the rose comb genes (see box 6 on page 93). However, if you look at boxes 8 and 14, you will note that a further 12.5% of the offspring were rose combed Lavenders, that carried single comb genes. These birds were indistinguishable from the pure rose combed lavenders because the rose comb gene is dominant. Overall, a total of 18.75% rose combed Lavenders were produced from the F1 offspring.

If you were just dealing with Black to Lavender and you wanted to work out the percentages of pure birds and carriers, it would be far easier and it works exactly the same as the rose to single combed cross of Wyandotte to Cochin:
Pure Black x Lavender = F1 100% Blacks carrying Lavender (visually Black)

F1 x F1 =
25% Pure Black
50% Black carrying Lavender (visually Black)
25% Lavender

F1 x Lavender =
50% Black carrying Lavender (visually Black)
50% Lavenders

F1 x Pure Black =
50% Pure Black
50% Black carrying Lavender (visually Black)
(There is little point in this cross as the Lavender carriers will be completely indistinguishable from the pure Blacks).

We've discussed Sibling matings in some depth and you should now have a good idea of how Dominant and Recessive genes recombine whether you are selecting for a single trait or two traits (which will reduce your percentages) as with the Lavender project.
I feel it's important to mention that trying to recombine more than 2 genes on a fowl, especially in pure form is like trying to hit a "home run" when you've never even played baseball. The chances of separate genes i.e. Lavender and rose comb recombining in pure form from an F1 mating are 1 in every 16 fowl produced. To add another factor to that slim chance would require the rearing of several hundred fowl. To once again use the Lavender Wyandotte project, Allan knew that the chances of producing yellow legged, rose combed Lavender birds from a sibling mating of the original cross (Lavender Orpington to Black Wyandotte) were very slim and so his objective was to first set the rose comb genes, secondly set the Lavender genes, at which point the chances of setting the yellow legged genes were increased greatly because he kept crossing back to Wyandotte. The moral in this example is to "set one trait at a time".
It can be quite a different case when working with Dominant genes. I always remember the late Dr Carefoot's example when he recreated Silver Pencilled Wyandotte Bantams in the UK. He brought the Silver gene in from Silver Brahma Bantams. Dr Carefoot realised (as anyone with a bit of sense would) that when working with Dominant genes in projects, it is sometimes possible to "kill 2 birds with 1 stone". In the case of the Silver-Pencilled Wyandotte, Dr Carefoot knew that by the time he purified the shape of his Wyandottes, most of the Brahma genes such as the feathered legs and Pea comb etc would be bred out. The Brahma was only used to bring in the Silver gene and hence the F1 and subsequent generations were bred back to Partridge Wyandottes until he produced some fine Silver Pencilled Wyandottes. His strains still remain today and have done a lot of winning over the years.

I realised a similar concept when I created the Pyle Wyandotte Bantams. I knew that the Large Pyle coloured female that I was using had terrible type for a Wyandotte, but by the time I had kept on breeding Pyle-looking birds to the Wyandotte Bantams, not only would the bantam size improve, but so would the type, and this is what happened.

The above examples prove that projects can be achieved in a faster time scale if a little research is done beforehand. If a little common sense is applied, you can achieve far better results. There have been many "counterintuitive" crosses over the years. People seem to fantasise about certain breeds, that when bred together, will produce particular colours - usually something new.

However, many people are on the right tracks with their theories but are soon put off by a "lack of results in the first cross". I recall one such project where someone was trying to create Lavender Wyandotte bantams. The person in question crossed a Lavender Pekin to a Black Wyandotte and was extremely despondent when all of the offspring were Black and so consequently gave up the project. However, if this person had done a little research, they would have realised that the offspring thought to be no good, held the key to the next step of the project (the Lavender genes).

The "What goes in, must come out" rule works to a large extent, and mainly with Dominant and Recessive genes. However it isn't always the case, as with the Blue gene for example. Say someone wished to recreate Blue-Partridge Wyandottes: they could breed a Partridge Wyandotte male to a Blue Wyandotte female and easily end up with 5 Black offspring. They might then think that carrying out a sibling mating from the F1 would yield a percentage of Blue Partridge birds. However, this wouldn't be the correct route in this particular example, because "Non Blue" birds do not carry Blue and hence cannot pass it on either.

The way forward in the Blue-Partridge Wyandotte project would be to breed a Partridge Wyandotte male to a Blue Wyandotte female and then cross any F1 Blue offspring back to the Partridge and follow this process until all of the self coloured fowl stopped emerging (whether Black or Blue) or until Blue-Partridge coloured birds could be selected. These could then be bred back to Partridge and would produce 50% Partridge and 50% Blue-Partridge. To determine the percentages of Blue-Partridge offspring (when Blue-Partridge is bred to Blue-Partridge), just apply the Mendellian principles of the Blue gene to the Black areas of the Partridge Wyandotte.

Genotype / Phenotype

These are complex sounding terms which denote whether or not a fowl is pure for the genes it appears to be pure for. So "Phenotype" means the way it looks. "Genotype" means it's actual genetic make up.
To give an example, if we crossed a pure Black Plymouth Rock to a pure Black Wyandotte, the F1 offspring would all be yellow legged Blacks sporting rose combs.
To those unaware of the genetic make up of the F1, they would just appear as Wyandottes with bad type.
Their Phenotype would be: yellow legged Black fowl with rose combs (because this is what they would appear to be).
Their Genotype would be: yellow legged Black fowl with rose combs carrying the single comb gene.
The Buff Orpingtons pictured bottom could each be carrying a copy of the Recessive White gene and hence would produce 25% White Offspring when bred together.

White Orpington males.

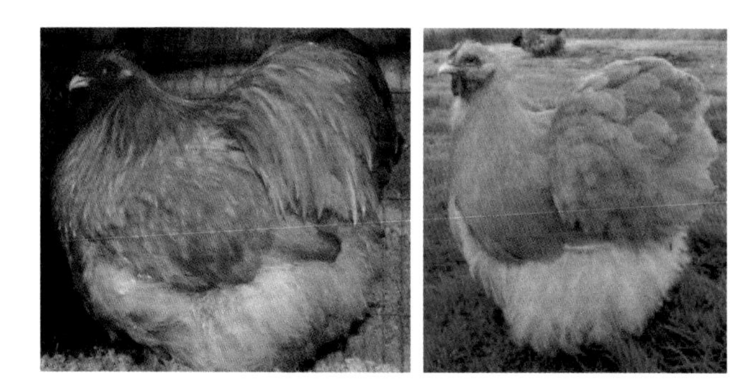

Their Phenotype is Buff Orpington, however their Genotype could well be Buff Orpington carrying one dose of Recessive White.

Making Use of Sex-Linkage

Sex-Linkage in poultry has been utilised for many decades and was the starting point of many modern hybrid egg layers. The most common cross was the Rhode Island Red male over Light Sussex females. This cross yielded two different colours of chick down - the females being Brown and the males being Yellow. Another common cross was that of the Rhode Island Red male over Barred Plymouth Rock females. This cross yielded chicks which were either Black, or Black with a Yellow spot on the head and lighter shanks. The Black chicks are commonly known as Black Rock Hybrids

The reason Sex-Linkage works is because a minority of genes are Sex-Linked meaning that the male can carry two copies - as with all other genes, but the female can only carry one copy, unlike with many of the other genes, where she can carry two copies. The genes which both male and female can carry two copies of are known as "autosomal".

In humans, males are XY and females are XX when it comes to chromosomes. However, in poultry it is the other way round and different gene symbols are used. Male chickens are ZZ and female chickens are Zw. That's easy. It is therefore easy to see why the female is the determiner of her offspring because males can only contribute male sex chromosomes to their offspring, whereas females can contribute either sex chromosome at a chance of 50%.

Sex-Linked genes are only carried on the Z chromosomes, and the most popular examples of such genes are: Silver, Barring, Bantam, Chocolate, and Slow Feathering.

Thus when a Rhode Island Red male is crossed to a Light Sussex female, he can only contribute Red genes to his daughters, however, as discussed the female is the determiner of the sex of her offspring, so if the Light Sussex female in question contributes the female chromosome (w) then the offspring becomes female (Zw) and is visually Red and consequently, Brown as a chick.

However, if she contributes the male chromosome (Z), then the offspring becomes male (Z/Z) - one male chromosome from each parent. Her son is visually Silver because his mother carried Silver on her male sex chromosome and since he requires that chromosome in order that he is male, he also involuntarily inherits the single copy of Silver that is carried on the same chromosome. Since Silver is dominant to Red (or Gold), then sons from

a Rhode Island Red male over Light Sussex females are visually Silver and consequently yellow as chicks.
It works exactly the same with the Rhode Island Red male over Barred Rock females. To demonstrate, I carried out a similar cross which was a Rhode Island Red male over Barred Wyandotte females. All females hatched Black with dark shanks and all the males hatched Black with a yellow spot on their head and with lighter shanks. Now you may wonder, as I did, how the females from a Red male and a Barred female could possibly be Black. This is because Barred Wyandottes and Barred Plymouth Rocks are genetically Black birds with the Barring genes added. Because the Barring gene is Sex-Linked, it was not possible for the female to pass it on to her daughters, only her sons (as with the Light Sussex female passing on the Silver gene to her sons). So, why aren't the daughters from a Rhode Island Red male crossed to Barred Rock or Wyandotte females, Red? This is simply because the Black genes which the Barred Rocks and Wyandottes carry are autosomal (not Sex-Linked) and are Dominant to the Red genes carried by the Rhode Island Red male.
The reason that sons of the Rhode Island Red over Barred Rocks or Wyandottes have lighter coloured shanks than their sisters is because the Barred varieties often carry many "Blackening" genes and usually one which extends to the shanks. However, the Barring gene is linked to a gene that cleans up Black pigment in shanks and so you will very rarely see a Barred or even Cuckoo fowl which displays any Black pigment its shanks. Females can display a bit of Black pigment in this area, however it is quite uncommon.

The above photos demonstrate the difference in Black pigment in the shanks of the offspring from a Rhode Island Red male over Barred Wyandotte females. Note the female is completely unbarred and how strong a single copy of the Barring gene is on the male.

Because Chocolate is a Sex-Linked Recessive gene and not Dominant, like Silver or Barring, then it is not visible in the male unless pure. So a male, that is a Chocolate carrier, irrespective of how much Black pigment he has, will not be identifiable to the naked eye. It is only from keeping records, or knowing that his mother was Chocolate that you can be assured he also carries the gene.

Let's look at Impures

What is an impure bird? The name implies quite simply that the bird in question does not carry genes in pure form and hence will not breed true to form when paired with a similar looking female. The term "Impure" when used in poultry breeding is often applied to males who, when all other genes are pure, carry a single copy of a particular gene. More often than not, this gene is Sex-Linked.

To use an example, if Barred Plymouth Rock Bantams were to be bred true in the UK, it would mean that the male of the variety would be lighter, more like the shade of a Barred Wyandotte. The true exhibition Barred Rocks are very similar in tone to the Barred Rock females. This is achieved by them only carrying a single dose of the Barring gene, and so in terms of Barring, Exhibition Barred Rock males are "Impure" for Barring.

Breeders of Barred Rocks have three main choices, they can breed a Light Barred male to Black females which will produce 100% Barred Offspring, the males being close to the Exhibition type. Or another option is to breed an Exhibition male to Exhibition females, which will result in the following:

25% Light Barred males - 2 copies of Barring

25% Dark Barred males (Exhibition type) - 1 copy of Barring

25% Barred females (Exhibition type) - 1 and only available copy of Barring

25% Black females - carrying no copies of Barring

The final option is to breed an Exhibition male to a Black female, this will produce:

25% Barred Rock males (Exhibition type) 1 copy of Barring

25% Barred Rock females (Exhibition type) - 1 and only available copy of Barring

25% Black males - carrying no copies of Barring

25% Black females - carrying no copies of Barring

As mentioned in the Cream section of "Changing Colours", Impure Silver Males are created by crossing Partridge and Silver Pencilled Wyandottes together - no matter which way round.

An Impure Silver male bred to a Silver Pencilled female will produce:
25% Impure males
25% Silver Pencilled males
25% Partridge females
25% Silver Pencilled females
An Impure Silver male bred to a Partridge female will produce:
25% Impure males
25% Partridge males
25% Silver Pencilled females
25% Partridge females
It is important to note that while the above percentages work, some breeders may experience the Cream gene emerging (Gold Duckwing effect), as has been the case in the past. Please refer to "The Cream Gene" under "Changing Colours" for information on how to test for the Cream gene.
It is not advisable to use Impure Silver males if you are wishing to seriously breed Silver Pencilled Wyandottes, as crossing to Partridge can sometimes introduce unwanted Red genes such as the unofficial but heavily tested autosomal Red genes, discovered by Brian Reeder with the proposed gene name of Ap (see "An introduction to Ap). However, there seem to be no detrimental effects when using Impure Silvers to produce Partridge males and females.

Heterozygous / Homozygous / Hemizygous

These scientific names sound very complex but Hetero, Homo, and Hemi are just Greek prefixes for the word "zygous". In basic terms they just refer to the amount of genes a bird is carrying at each allele (the two Loci where a gene can be carried in single or pure form).
To give an example, a Splash bird is "Homozygous" for the Blue gene, because it carries two copies. A Blue bird is "Heterozygous" for the Blue gene because it only carries one of the two copies that it could carry.
The word "Hemizygous" refers to the Sex chromosomes, in particular, females, where they can only carry 1 copy of a gene at the most, and on the long chromosome (Z). To use an example, Barred females are "Hemizygous" for barring and so in scientific terms are written as "B/_". The underscore refers to the fact that Barring cannot be carried on the short, female chromosome (w).

Understanding the + side

Pure Barred males would be written in scientific terms as "B/B". Impure Barred males would be written as "B/b+". But where does the "b+" come from and what does it mean? I know that "B" stands for Barring, but why make it look complex by having a plus symbol? The reason is simple. All the plus stands for is the location where Barring could be if the bird in question was mated to a Barred bird. With every imaginable colour or feature that is different from the Red Jungle Fowl - assumed by many to be the original and unaltered form of fowl - there has to be a location where these genes could be carried. Every gene in the Red Jungle Fowl is marked with a plus symbol by geneticists because it is completely unaltered by any genes such as Blue, or White, or Leg feathering. The use of a capital letter or otherwise and "plus symbol" just indicates the normal alternative for the gene in question, (whether dominant or recessive to the mutation) for example, if a rose comb bird is R/R, then a single comb bird is r+/r+.

An Introduction to "Ap"

Brian Reeder of Kentucky, USA, observed that not only is the Silver gene unable to change the breast colour of Black-Red females and the back colour of Wheaten females, but the residual and "unaltered by the Silver gene", Red pigment was able to be manipulated. Brian carried out his test matings over a 10 year period and he published his findings in his book "An Introduction to the Colour Forms of Domestic Fowl".

Brian suggested the official naming of the gene to be "Ap" standing for "Autosomal Phaeomelanin". This in basic terms means: Non Sex Linked Red, so a form of Red pigment that is basically unaltered by the Silver gene - unlike Sex-Linked Red, which is altered by the Silver gene and makes a Partridge Wyandotte into a Silver Pencilled Wyandotte.

It is clear when observing Salmon Faverolles that the Silver gene has little effect on Ap. Red enhancers such as Mahogany can saturate very well in the shoulder and back areas of the male, and across the whole of the back area of females.

Brian hypothesised that fowl were either Ap in pure or single form, or were completely devoid of the gene. In Brian's view, the pure white Silver birds, such as Silver Pencilled Wyandottes have to lack Ap in order to be free of rust or any hint of Red tone whatsoever.

He carried out numerous test matings and came to the conclusion that his theory was correct and that it was possible to manipulate the Ap gene and breed it either into or out of a strain, depending on what was desired by the breeder.

Brian has yet to officially publish his findings, however it is sufficient to say that his theories are very plausible and account for a lot of what is going on in poultry plumage genetics. I am a firm believer in his proposed gene and have made much use of the knowledge of it in many of my breeding programmes. Brian is owed a lot of credit for dedicating so many years to his research which he carried out to an exhaustive extent, not only with the Ap gene, but with many other factors covered in his book.

The above photos show a Silver Grey Dorking female (top left) possessing two copies of the Ap gene, a Gold Phoenix female (top right) possessing only a single copy of the Ap gene, and a Silver Phoenix female devoid of the Ap genes.

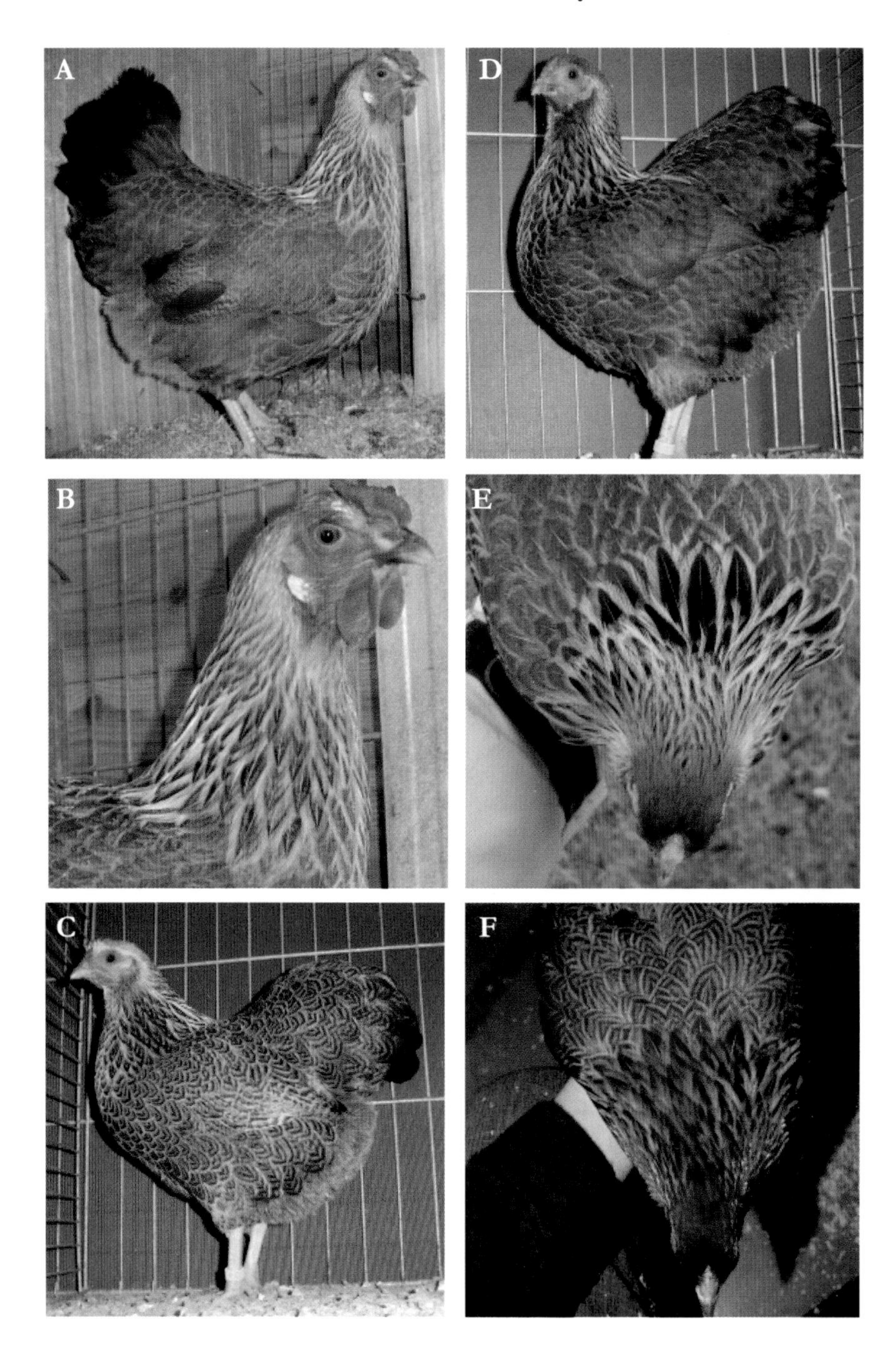
A
D
B
E
C
F

In my opinion, the top left photo (Picture A) is the most convincing proof to date that the presence of the proposed "Hackle Black" (Hb) gene is autosomal and segregates independently of eb. It has been assumed (with great plausibility) that females which are unpatterned "eb" at the "e" locus (such as Partridge Exhibition cock breeding females) display Dark Hackles as standard, and that adding the Pattern gene (Pg) is enough to "Pencil" the necks of such females. However, with close observation, it would appear that adding the Pattern gene (Pg) to an eb female which displays the proposed Hackle Black has little effect on the neck area (Picture F) which remains Dark. Although the birds on the far left (pictures A, B and C) have the Cream gene (ig), the sequences - A to C & D to F from top to bottom show the effects of adding the Pattern gene (Pg) to first a "non carrier" of Hackle Black (A+B), and second, a Hackle Black carrier (E+F). Well pencilled birds with good neck pencilling should be devoid of the Hackle Black genes, proposed as "hb+", like the bird in the bottom left picture.

It is also my opinion that the Hackle Black genes don't only Blacken the neck and saddle hackle striping in fowl, but have another role in darkening the whole body, and making Black Breasted males possible even when the Pattern gene is present in pure form. This thinking is further reinforced in females, an example being the Gold Brahma. If you compare a Gold Brahma female to a Partridge Wyandotte female, the difference in thickness of pencilling is quite evident in most cases, which, I believe is a result of the presence or absence of the Hackle Black gene.

One observation that led me to believe the Pattern gene could not really alter the effect of Hackle Black in Exhibition males (such as the Partridge Wyandotte), was that German lines seemed to carry the Hackle Black genes and be devoid of the Pattern gene to produce Exhibition males, whereas English strains seemed to carry the Hackle Black genes as well as the Pattern gene with exactly the same effect in the Black areas of the Exhibition male. This suggests that Hackle Black genes are absolutely essential if one intends to breed Exhibition male lines, and perhaps even more importantly, recreate the Large Partridge Exhibition males, which are thought have been extinct in the UK for many years.

Chocolate Extra

In my opinion, the Chocolate gene and what is currently known about it is fairly limited. While Dr Carefoot discovered the gene, and that it operated as a Sex-Linked Recessive as far back as 1994, very little research has thus far been carried out to determine what other effects the chocolate gene may have in poultry, such as whether it dilutes Red pigment, its effects on feather growth rate, whether it is consistent in expression, or rather variable like the Blue gene, and perhaps more intriguingly, how Chocolate and Blue would look on the same bird?

Fortunately specific test matings were set up in the hope of answering some of the above queries. I have always stood by my personal beliefs that the Blue gene doesn't have "Variable Expression" but rather is effected largely by how much Black there is in the bird in question. In my opinion, the expression of Chocolate works in much the same way as the Blue gene, meaning the more Black present, the darker the shade of Chocolate.

A "Chocolate carrying" Black Orpington male was crossed to a Partridge Wyandotte female. All of the F1 male offspring were Black, and half of the F1 female offspring were Black; the remaining female offspring were Chocolate (25%) which was consistent with expectations. However, it was noteworthy that even though the F1 offspring were genetically half Partridge, the Black genes they carried were very Dominant as expected, all Blacks were a glossy "Beetle Green" shade and all Chocolates appeared as "Chocolate" versions of the Blacks, being equally as Glossy.

With this is mind, I carried out sibling matings of the F1 offspring to purposely breed off some of the Dominant factors responsible for a "Glossy" Black plumage, in an attempt to produce some "Matt Black" birds and hopefully, some "Matt Chocolates" which would prove that any difference in the particular shade of Chocolate in any fowl is fully reliant on the quantity of Black pigment available.

The above test matings were very successful, and of the 27 chicks produced, there were 6 "Matt" fowl (3 Matt Blacks and 3 Matt Chocolates) as well as the expected percentages of Glossy Blacks and Glossy Chocolates.

Glossy Chocolate.

Matt Chocolate.

Glossy Black.

Matt Black.

The above photos depict the difference in tone between Glossy and Matt fowl whether Black or Chocolate: the Glossy birds are to the left and their respective Matt alternatives are to the right. I don't believe that the differences in expression of Black and Chocolate are due to the presence or absence of the Sex-Linked Gold or Silver genes, as the Black Orpingtons were tested to be genetically Gold, ruling out the possibility of the Silver gene having any effect in the aforementioned experiment.

Pigment Diluting Effects of Chocolate

It is believed by some that the Blue gene has a slight "diluting" effect on any residual Red pigment in a fowl. However, not everyone agrees, so what about Chocolate?

The above feather of the two pictured below was plucked from the neck hackle of a known "Chocolate carrying" Partridge Wyandotte Bantam male. The feather was the only one of its kind in the bird's plumage and the one below is synonymous with rest of the neck hackle feathers.

On close inspection, it would appear that the Chocolate gene has a diluting effect on Red pigment, which if you observe, gets weaker towards the distal end of the feather, perhaps coincidentally at the same point that the Chocolate gets weaker?

The Jury's still out as to whether the Chocolate gene dilutes Red pigment, and much more proof will have to be gathered before any conclusions can be drawn. However, it is a distinct possibility at the present time.

Chocolate and Blue in the same bird

This was an interesting prospect and one which could only be determined by carrying out specific test matings to provide the answers.

In order to guarantee producing offspring which would each carry a single copy of both the Blue and the Chocolate genes, it was necessary to cross a Chocolate Orpington male with a genetically "Blue Splash" Wyandotte female.This way, I knew that all females produced would have to display the effects, if any, of combining the Chocolate and Blue genes because they inherited a Chocolate gene from their father and a Blue gene from their mother. The effect was quite interesting and could be described as "Light Chocolate" but not "Khaki." The males produced from this first cross were Blue, as expected because they carried only a single dose of Chocolate.

A Chocolate male paired with a 'Blue Splash' female.

The resulant 'Mauve' looking chick.

The Chocolate gene and Feather Growth Rate

It was observed that Chocolate Orpingtons feathered more slowly than the already 'slow feathering' Black Orpingtons and that the Chocolate gene appeared to have an "Epistatic" effect on feather growth rate in general. "Epistasis" is where a gene that has a particular function is linked to another gene. An example is the Barring gene where it reduces shank pigment as well as "Barring" feathers.

Note how the Partridge Wyandotte female chick to the right is feathering much more quickly than her Chocolate-Partridge sister to the left of her. In this mating, the Chocolate-Partridges were all female, so the possibility of sexual differences having an effect was ruled out. Results were also consistent in future generations, where all Chocolate, or Chocolate-Partridge chicks feathered up much more slowly than their respective Non-Chocolate relations.

Sex-Linked Recessive Chocolate - not to be confused with other similar looking genes

Black Minorcas.

Partial Albino Minorcas.

Shortly after the new Millennium, an extremely rare genetic mutation showed up in a line of Black Minorca Bantams owned by Mr Allan Brooker of Sandhurst in the UK.

Allan noted that approximately 25% of the chicks produced from a particular breeding pair of Black Minorcas were quite different from the rest. He could tell from day old that these chicks were different because they displayed Pinkish down and leg colour, and had distinct Pink eyes. When they feathered up, they appeared to be very similar in plumage to birds displaying the Sex-Linked Recessive Chocolate gene, discovered by Dr Carefoot.

However, because these Pink eyed birds turned out to be males and females, the possibility of "Sex-Linkage" was ruled out, leaving the likely alternative of the causative gene being an "Autosomal" Recessive (Non Sex-Linked Recessive). The effect is likely similar to the "TRP-1" (Tyrosinase Positive) Albinism gene found in mammals and avian species. In basic terms, the Black Pigment production process is severely hampered which results in a lack of Black pigment in the eye and shanks, although some is allowed to come through to the plumage which is why it is classed as "Tyrosinase Positive." If the birds in question were white in plumage as well as displaying pink eyes and white shanks, it is likely that "Tyrosinase Negative" albinism would be present. Some refer to "Chocolate" Minorcas as carrying the "Pink eyed Dilute" gene.

The Dun Gene

The Dun gene is very similar to Chocolate and is long established in poultry, particularly in the Game varieties. Another breed to display the Dun gene is a fowl called the "Vogtlander" of German origin, pictured below right.
Dun operates very similarly to Blue in that it doesn't breed true and is an "Incomplete Dominant". Its effects are caused by the presence of a single dose, and the effect of two doses is a very light, almost cream shade known as "Khaki."

The Polands pictured above show the difference, compared with a Black Poland (top middle) when a single copy of the Dun gene is added (top left).
The bird above left is "Khaki" (two copies of Dun).
Pictured to the right is a "Khaki" (Dun Splash) chick and a Dun chick to the right of it.

A Red Wyandotte male, sometimes mistaken for Chocolate.

A Red Wyandotte female, sometimes mistaken for Chocolate.

Mottling Complexity

Mottled varieties of poultry, including Mille Fleur Belgians, Speckled Sussex and Ancona to name but a few, all rely on the "Mottling" gene (mo) reported as recessive. It works by inhibiting all pigment at the distal end of the feather, and then producing a Black band around it, before normal pigmentation of the variety in question resumes. However, it is a complex gene and little research has been carried out to determine the expression of mottling. In my opinion, the expression of mottling is largely dependent on the purity of the plumage genes of each respective mottled bird. If the correct genes are set, then the chances are that the mottling will be far more concise; an example is the Ancona.

Dr Carefoot concluded that the pattern of the Exchequer Leghorn was caused by the presence of the mottling gene (mo), and crosses to Anconas resulted in intermediate offspring between the two varieties. This suggested that the mottling gene (mo) is responsible for the mottling in all mottled breeds, however its expression is altered greatly by hormones and the presence of other genes. Even and precise mottling requires rigorous selection, but can be fixed to a point within a strain.

It is my opinion that the difficulty in achieving mottling on the neck of many otherwise "Black mottled" varieties such as the Wyandotte, Pekin, and Ancona, is attributed to the presence of more than the minimum required number of genes to make the bird Black. Test matings would suggest that the removal of such excess Black genes, allows for normal mottling to resume in the neck area of any given fowl.

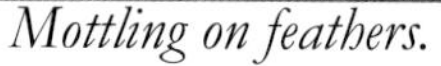

Mottling on feathers.

Black Mottled Pekin.

Ancona Bantams.

Spangled Wyandottes.

Hen Feathering

The vast array of colours and patterns seen in poultry today led me to a specific question: why do some breeds have "hen feathered" varieties, meaning that the male displays exactly the same feathering as his female counterpart, when others do not?
To name a few examples: breeds that have "hen feathered" varieties include Hamburghs, Game, Laced Wyandottes (in the US), and Sebrights and Campines are required to have "hen feathered" males as standard.
The gene responsible for hen feathering in males was assigned the gene symbol "Hf" (short for hen feathering), the capital H indicative of its dominance over normal male feathering. It works by increasing the amount of aromatase production in males and the result is more oestrogen in the feather folicles, and hence hen feathers.
I was curious to know whether "hen feathering" was only possible on certain plumage patterns, or whether as suspected, the gene would have the ability to alter the expression of male feathering to female feathering, by increasing the amount of oestrogen in males from any given breed or variety.
I was intrigued to see whether Pencilling could be achieved on Silver Pencilled Males throughout the whole body. The aim was not to change the standard, nor introduce a new form of "pullet breeding" Partridge or Silver Pencilled male. The aim was simply to see what could be done with the hen feathering gene.
Dr Clive Carefoot discovered that the pattern of Silver Pencilled Hamburghs and Silver Pencilled Wyandottes was separated only by a single gene, Dark Brown (Db). With this in mind, crosses were made between Silver Pencilled Wyandotte bantams and "pullet breeding" Gold Pencilled Hamburghs; Silvers proved difficult to locate. Resulting backcrosses would suggest that a single dose of the hen feathering gene is enough to alter the expression of the usual pullet breeding Pencilled Wyandotte male plumage from "male feathered" to "hen feathered."
Until it can be proven otherwise, this hypothesis, and strong supportive evidence, would suggest that the presence of the "hen feathering" gene could alter any male plumage pattern to that of its respective female counterpart, even in single dose, but especially in pure form (two doses).
The male opposite top left is a direct cross between a Gold Pencilled Hamburgh

male line (normal feathered males) and a Silver Pencilled Wyandotte line, suggesting that the single presence of the Dark brown gene (Db) is recessive in appearance. The male above right is genetically very similar with the addition of only a single dose of the Hen feathering gene.

Dutch Silver Pencilled 'pullet breeding' males with normal feathering.

Possibilities with Lacing

When I first joined the Laced Wyandotte Club several years ago, there was discussion over whether additional Laced varieties could be produced aside from the established Silver Laced, Gold Laced, Buff Laced and Blue Laced. Since then, Blue Laced Silvers, sometimes called "Violet Laced", have been produced in Large Fowl and Bantam (see photos), the Large Fowl by Richard Davies, and the Bantams by Clare Skelton.

Silver, Blue, Buff, and Blue Laced Silvers, all rely on altering one or both of the Black and Red pigments which would otherwise give the effect of "Gold Laced", and aside from adding Chocolate to that equation (giving Chocolate Laced Silvers or Reds), there would appear to be very few other possibilities with Lacing.

However, if we consider the above are all made by altering the Gold Laced, which consists of a certain genetic make up, we can look to other possibilities such as the Silver Sussex, and how altering the shades of that pattern will open the door to many other possibilities.

The best Silver Sussex females are well Laced and this pattern could be introduced to the Wyandotte, not only allowing the possibility of White Laced Blacks, as are Silver Sussex, but all the possible combinations of altering the Black ground colour or the White Lacing i.e. White Laced Blues, Red Laced Chocolates, Red Laced Blacks etc.

Buff Laced in their true form are Gold Laced Wyandottes with the addition of one or two copies of the Dominant White gene. However, the Splash Laced Reds which emerge from Blue Laced Reds can resemble the Buff Laced (or White Laced Reds) and many have been shown over the years, especially in the UK. They are not correct because the residual Blue splashes spoil the Hackle.

It is claimed that Buff Laced fowl on the Continent breed true whereas the UK strains which rely on the Dominant White gene do not. A possible answer to this anomaly is that Dutch strains have fixed in them the Mahogany gene and likely the Ap gene, which in combination counteract the strength of the Dominant White gene in pure form, thus not allowing any "pales" (almost White birds) to be produced.

Blue Laced Wyandotte.

Blue Laced Silver Wyandotte.

Gold Laced Wyandottes.

White Laced Reds with dominant White (I).

White Laced Reds with purity of Blue (BI).

Silver Laced Wyandotte females.

A Prototype Chocolate Laced Red.

The Importance of understanding 'Gene Interaction'

It is important to understand how certain genes interact with each other and how some effects are produced by the presence of more than one factor. It doesn't always stand to reason, and the combinations of some genes produce patterns that are perplexing to the genetics student. One such combination is that of the Gold Spangled Hamburgh. Genetically, it is exactly the same as the Gold Pencilled Hamburgh with the addition of the Black gene, Melanotic (Ml).

A Gold Pencilled Hamburgh.

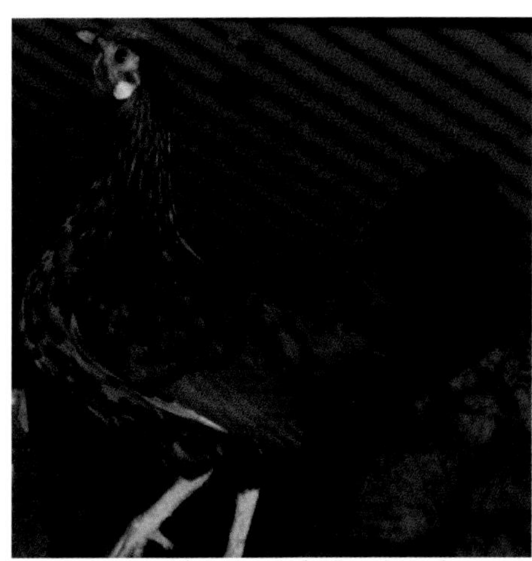

A Gold Spangled Hamburgh.

If we look at the effects of the Melanotic gene on a Partridge Wyandotte female, it is clear to see the gene's effect of "Blackening" the Pencilling (Barnevelder female), so Blackening the Black parts of the fowl. Adding the same gene to a Gold Pencilled Hamburgh produces Spangling (Carefoot,1985) which is confusing, because one would expect merely a heavier "Hamburgh Pencilling." This is why it's important to understand that not everything in poultry genetics is clear cut and that the laws of "Interaction" must be factored in to breeding programmes.

Something Lacking?

In order to understand how all the different plumage colours and patterns are made up, it's essential to realise that each colour or pattern is only possible because of the genes each bird "lacks" as well as the genes it carries.
This may seem really obvious, but to explain a bit more clearly, a petrol engine wouldn't function if diesel was put in the tank as well as petrol. A petrol engine needs to be completely devoid of diesel for the engine to function.
Varieties are formed because the genes they possess "click" to give a colour or pattern that is pleasing to the eye. To give an example, the beautiful Mahogany Rhode Island Red colour is produced by the purity of the Red intensifying gene, Mahogany (Mh) on an extended Red Wheaten background. If the bird in question was also carrying Dilute (Di), it would look more like a New Hampshire Red, so that is why it's important for birds to "lack" certain genes as well as carry the appropriate ones essential to the appearance of each respective variety.

A Pair of Serama Bantams which in most cases still segregate many colour genes.

Genetic Anomalies

Every so often, a colour or pattern emerges which doesn't "fit" with the current accepted genetic knowledge. One such pattern is the "Braungebandert", a German name for a colour of Wyandotte that seems quite unique to them. Males are very similar in appearance to Rhode Island Reds, but have a dark under colour to the feathers. They are a rich Cherry Red all over, with the shoulder feathering being slightly darker than the rest of the bird. Females are at first glance quite similar to Exhibition Partridge Wyandotte females; however, on closer inspection, their ground colour is seen to be darker and is closer to Brown than Gold. The other major difference in the female is the evident lack of multiple pencilling on the breast progressing up towards the neck. Instead of finely pencilled feathers, the Braungebandert females have in most cases, a single concentric Black ring on Brown feathers in the upper breast region.

To give an insight, females which are "Partridge" pencilled, no matter to what degree, are usually paired with and produce, males that display predominantly Black breast, thigh and central wing feathering (the bit between the shoulder and the wing bar), as well as having mainly Black flight feathers. Thus the Braungebanderts have perplexed many. Every time a possible explanation for their make up is put forward something always contradicts it, and many people, myself included, have ended up back at square one.

One proposed explanation was the presence of the "Mahogany" (Mh) gene but that idea was soon discredited by the example of "American Partridge Wyandottes" which are finely pencilled and very dark. The males display a rich Mahogany colour with Black Breasts, suggesting that Mahogany was not the gene responsible for extending Red pigment to such a degree. Other genetics enthusiasts have proposed that the gene responsible for the Braungebandert is Dark Brown (Db), but again that idea was soon discredited by the work of Dr Carefoot who felt strongly that the presence of Dark Brown on a Partridge bird would cause "Hamburgh Pencilling" (transversely Barred feathers).

In my view, the Mahogany gene can be ruled out at present because if it had the strength to extend Red pigment to a large degree on the breast area of males, then varieties such as the Salmon Faverolle wouldn't be possible because males would have Red feathers in the usual Black feathered areas.

It is not beyond the realms of possibility that a mutation exists in Germany that is currently unknown in the rest of the world; or it could be that the Red extending gene responsible for the Braungebandert already exists outside Germany, but is disguised in combination with other genes. One possible explanation is a gene in Barnevelders which allows for some males to have "Laced" breasts, suggesting that the removal of the two genes partially responsible for such lacing (Pattern gene Pg and Melanotic Ml) would yield males very similar in appearance to that of the Braungebandert male. However the laced breasts of some Barnevelder males could just be caused by the presence of the partial lacing unit plus Mahogany.

I feel that the most likely reason for the Braungebandert appearance is the presence of an unknown mild Red extending gene added to a Partridge "Pullet Breeding" colour base, which obviously has more effect on males than females. The gene name is proposed as "Bourbon" with the possible abbreviation of "Bor". However at present this thinking is only hypothetical and may be proved incorrect in time.

A Braungebandert female.

A Braungebandert male.

Pictured Below are Braungebandert males and females displaying the unquantified and unknown Red pigment extending gene(s), which I propose as "Bourbon" (Bor).

A Braungebandert male.

The view of the breast of a Braungebandert female.

An American Mahogany Partridge male.

An American Mahogany Partridge female.

Genetics of Egg Shell Colour

The genes responsible for the great many shades of egg colour are some of the most difficult genes to quantify. White eggs are the result of a "lack" of pigment whereas tinted and the several shades of darker brown eggs are caused by the amount of "Protoporphyrin" pigment present.

The genes which regulate Protoporphyrin levels are thought by geneticists to be "Polygenic", meaning there are several factors at work. Sex-Linked genes are thought to have some involvement, and a Sex-Linked recessive removing all pigment was reported by Shofner (1982), noted as "pr".

Blue or Green eggs are the result of an Autosomal Dominant gene termed "O" described by Punnett (1933). The actual shade of Blue or Green eggs respectively is highly dependent on other factors such as the levels of Protoporphyrin in the female. Blue eggs are a result of the presence of the Blue egg gene "O" and the lack of Protoporphyrin, whereas Green or Olive coloured eggs are the result of the "O" gene with clear presence of the Protoporhyrin pigment.

Just as the range of shades from tinted through to dark brown egg shells is vast, the range of available shades of Blue / Green shells is just as wide, obviously affected greatly by the lack or presence of Protoporphyrin pigment and the relative quantities.

The one major difference that the Blue egg gene has compared with Brown or White eggs is the fact that it permeates the shell and thus can be seen from the inside. Brown pigment is deposited on the outer shell and can be rubbed off with a fair degree of ease.

The general consensus is that to improve the colour of a dark egg strain, you should always breed from the females who lay the darkest eggs (and their sons). The shade of dark brown pigment usually is a lot darker when a pullet begins laying than it is when she is approaching the annual moult. This is no doubt a result of the change in hormone levels.

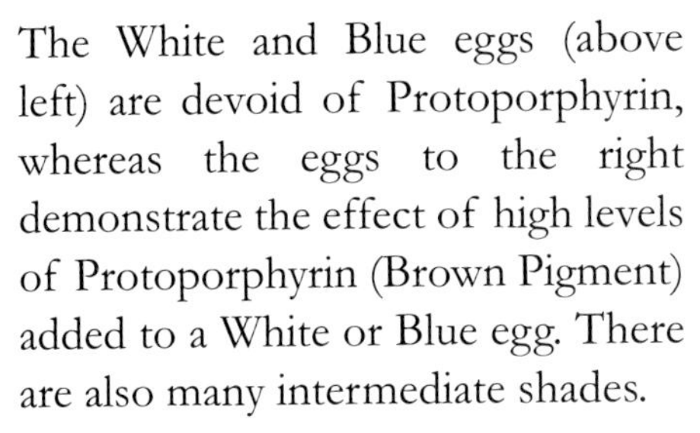

The White and Blue eggs (above left) are devoid of Protoporphyrin, whereas the eggs to the right demonstrate the effect of high levels of Protoporphyrin (Brown Pigment) added to a White or Blue egg. There are also many intermediate shades.

Above: the varying shades of Blue eggs from an Araucana breeder. It was noted that while the Blue egg gene (O) is Dominant to the non Blue egg genes (white), carriers of a single dose produced eggs with a slightly lighter Blue tone than those which carried the gene in pure form.

Wheaten Marans.

Eggs from Wheaten Marans.

White Leghorns.

Eggs from White Leghorns.

Araucanas.

Eggs from Araucanas.

Light Sussex.

Eggs from Light Sussex.

Lemon-Cuckoo Pekin male.

Lemon-Cuckoo Wyandotte female.

The relatively new "Lemon-Cuckoo" colour is becoming increasingly popular. It was introduced into Cochins by Jack Killeen of Yorkshire, but breeders of Cochins have reported difficulties in obtaining Lemon-Cuckoo females. In my view, the principles of Black and White Barring (dosage effect) apply to Lemon-Cuckoo and in theory, breeding a Light Lemon-Cuckoo male to a Buff female should result in 100% Lemon-Cuckoo Offspring. If a Lemon-Cuckoo male is only carrying a single dose of the "Sex-Linked" Barring gene (as is the one above), then half of his daughters will be self Buff no matter what his partner in such a cross carries. This is just something to keep in mind when working with this colour.

Frizzling Factors

A Frizzled Cochin male.

A White Frizzle female.

Breeders of closed flocks of Frizzled varieties often report the emergence of one quarter normal feathered, and one quarter "over Frizzled," as well as 50% "Frizzled" offspring when exhibition Frizzles are bred together. The Frizzled effect is caused by the presence of a single dose of the Frizzle gene (F) discovered by Hutt 1930. Birds carrying two copies of the "F" gene tend to be over Frizzled. The Frizzle gene fits well with Mendel's law of segregation, and breeding an over Frizzled bird to a normal feathered bird (from the same closed flock) should produce 100% Exhibition Frizzled offspring. However, Hutt also discovered a supressor of the gene "fm" (Frizzle modifier) which actually supresses the effects of the Frizzle gene. This gene needs to be absent in Exhibition flocks.

Sometimes birds can appear Frizzled, however the effect is usually caused by a malfunction of the preen gland or anaemia, rather than the well documented Frizzle gene.

The bottom feather (of the 3 above) is the from the neck hackle of a Partridge Wyandotte male. The middle feather shows the effect of adding a single copy of Recessive White (Lemon), and the top feather shows the effect of 2 copies of Recessive White.

The above feathers show the effect of the Lavender gene in pure form on a Black feather.

The above feathers show the effect of adding the Silver gene in pure form (2 doses) to the neck hackle of a Partridge Wyandotte male. The effect is Silver Pencilled.

The above feathers show how the Sex-Linked Barring gene affects Black Pigment by repeatedly interrupting and resuming the Pigmentation process. The Barring gene is effective on any colour or pattern, not just Black.

The sequence above shows the effect of adding first a single copy of the Blue gene (middle) to a black feather giving the effect of Blue, and second, adding 2 copies of the Blue gene to a black feather, producing the "Splash" effect.

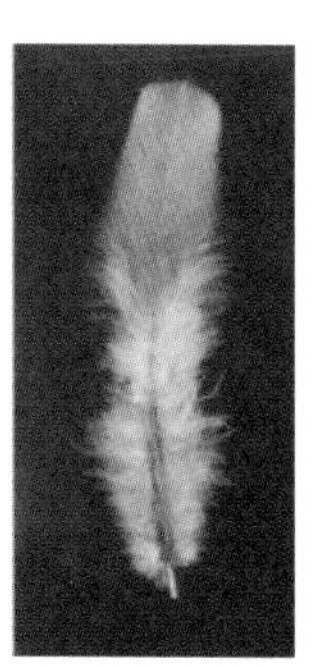

The above feather sequence shows the effect of adding the Dun gene in single dose to a Black feather (middle), and to the right, adding two copies of the Dun gene to a Black feather (Khaki or Dun Splash).

The above feathers show the effect of the Sex-Linked Recessive chocolate gene on a Black feather.

The lower feather from the 2 above is from the neck hackle of a Partridge Wyandotte male. The feather above shows the effect of adding a single dose of the Dominant White gene.

The above feathers show the effect of adding the Pattern gene in pure form (2 doses) to a "cock breeding" (eb) Partridge Wyandotte female.

The above feathers show the effect of adding the Ginger gene to a pencilled feather. The Ginger gene has many effects when combined with other genes; this is its effect on the feather of a Partridge Wyandotte hen.

The above feathers show the effect of the Columbian gene on a genetically brown (eb) female. The Columbian gene has the same effect on both males and females and in pure form, overrides the pattern gene (Pg) in the absence of the Melanotic (Ml) gene.

The above feathers show the effect of adding the Melanotic gene together with the Pattern gene to a Columbian Restricted feather. This is the same as adding the Columbian gene to the body feather of a double laced Barnevelder: the effect is Single Laced.

Some Popular Genes

To list fully the number of genes in poultry is beyond the scope of this book; it may be the subject of another book. However, I have compiled what I feel are the most popular genes and have given a short description of the way they work with other genes and whether they are regarded as Dominant, Incomplete Dominant, or Recessive.

The Blue Gene (Bl) - BATESON & PUNNET (1906)

The Blue gene alters the shape of the usually circular Black pigment granules within the feather and hence they become oval shaped and distorted, giving the appearance of Blue colouring.
Classification: Incomplete Dominant.
Breeds true? X (Except the Splash versions)
Examples of breeds with Bluc: Blue / Buff Laced Wyandottes, Blue / Splash Orpingtons, Blue Andalusians.

The Dominant White Gene (I) - BATESON (1902)

The Dominant White gene inhibits Black Pigment from coming to the feather, but in single form has little effect on Red pigment, hence the possibility of Pyle.
Classification: Incomplete Dominant.
Breeds true? √ To a point, however in pure form the Dominant White gene has a strong effect of diluting Red Pigment unless any Red intensifiers such as Mahogany are present. The way the Dominant White gene operates is still not fully agreed on by geneticists. However, in combination with the Ginger gene, the Dominant White gene has been proven to breed true, the best example being the Chamois Friesian Fowl.
Examples of breeds with Dominant White: Pyle Game, Buff Laced Wyandottes, Chamois Friesian Fowl, Chamois Polands, White Leghorns.

The Lavender Gene (lav) - BRUMBAUGH (1972)

The Lavender gene restricts the transfer of Black pigment to the feather and hence a pastel shade of Lavender is produced. The gene does breed true, however it is sometimes linked to a gene that causes very rough tail feathering. Its effect on Red Pigment is to dilute it heavily to a straw colour.
Classification: Recessive
Breeds True? √
Examples of Breeds with Lavender: Araucanas, Pekins, Leghorns.

The Silver Gene (S) - STURTEVANT (1912)

The Silver gene inhibits Red pigment from coming to the feather and is the gene necessary for all Silver varieties of fowl.
Classification: Sex-Linked Dominant
Breeds True? √
Examples of Breeds with Silver: Pencilled Wyandottes, Light Sussex, Barred Rocks.

The Pattern Gene (Pg) - CAREFOOT (1985)

The Pattern gene arranges Black pigment in an order that depends largely on the presence or absence of other genes. On a partridge Wyandotte "Cock-breeder" female, it turns the mossiness into concentric pencilling. It is an important factor in "Hamburgh pencilling" (transverse bars) and plays a part in the Spangling and Lacing patterns.
Classification: Dominant
Breeds True? √
Examples of Breeds with the Pattern gene: Partridge Wyandottes, Spangled Hamburghs, Sebrights.

The Melanotic Gene (Ml) - MOORE & SMYTH (1971)

The Melanotic gene is recognised for its "Blackening" properties. It makes concentric pencilling into heavy double lacing. It is found in many Black fowl and in many cases helps to make a solid Black bird.
Classification: Dominant
Breeds True? √
Examples of Breeds with Melanotic: Barnevelders, Black Wyandottes, Spangled Hamburghs, Indian Game, Laced Wyandottes.

The Columbian Gene (Co) - SMITH & SOMES (1965)

The Columbian gene restricts Black pigment and allows for more Red pigment in a fowl, except for the primary, neck, and tail feather areas where Black Pigment is largely unaffected. Added to a Double Laced Barnevelder, it produces the Lacing effect, as seen in Laced Wyandottes.
Classification: Dominant
Breeds True? √
Examples of Breeds with Columbian: Columbian Wyandottes, Laced Wyandottes, Lakenvelders, Vorwerks, Light Sussex, Buff fowl.

The Ginger Gene: Officially Dark Brown (Db) - MOORE & SMYTH (1972)

The Ginger gene is similar in some ways to the Columbian gene, however it has a stronger effect on neck hackles, particularly in males, allowing less Black pigment and more Red pigment. It has a role within many patterned varieties.
Classification: Generally Recessive, but is known to vary in expression
Breeds True? √
Examples of Breeds with Ginger: Pencilled & Spangled Hamburghs, Ginger Oxford Old English Game, Friesian Fowl, Gold Sebrights, Buff fowl.

The Barring Gene (B) - SPILLMAN (1908)

The Barring gene's main function is to interrupt the pigmentation process. This can be clearly seen in Barred fowl where the Black pigment is allowed to come to the feather for a time and then stopped, and then allowed to continue further up. This process is repeated along each feather from the distal end to the base. Precise Barring such as that of the Plymouth Rock relies on the presence of the Slow Feathering gene (Ks) without which it would be more like Cuckoo. The Barring gene doesn't occur only in Barred Fowl; it is found in other varieties such as Crele and Buff-Cuckoo. It relies heavily on minor modifiers which when combined, make for a better effect.
Classification: Sex-Linked Dominant
Breeds True? √
Examples of Breeds with Barring: Barred Rocks & Wyandottes, Crele Game, Legbars, Buff-Cuckoo Pekins, Cuckoo Marans.

The Chocolate Gene (choc) - CAREFOOT 1995

The Chocolate gene interferes with the production of Black pigment. It is very similar in appearance to a gene called "Dun". However, it breeds quite differently and has only been known in the UK in Orpingtons since 1994. It has little or no effect on Red pigment.
Classification: Sex-Linked Recessive
Breeds True? √
Examples of Breeds with Chocolate: Orpingtons, Seramas, Wyandottes.

The Dun Gene (I^D) - ZIEHL & HOLLANDER (1987)

The Dun gene, sometimes called "Fawn" or "Chocolate", works in much the same way as the Blue gene. However, its expression is a Chocolate shade in single dose and "Khaki" in pure form. Its effects are caused by the reduction of amino acids in the Black pigment formation process.
Classification: Incomplete Dominant
Breeds True? X (Except the Khaki versions)
Examples of Breeds with Dun: Old English Game, Polands, Wyandottes (US).

The Recessive White Gene (c) BATESON & PUNNETT (1906)

The Recessive White genc inhibits both Black and Red Pigment from coming to the feather and hence produces an all White appearance. There are a further 2 forms of Recessive White, with "c" being the most Dominant of the three. Pure Recessive White birds can sometimes display "bleeding" which is a faint Red pigment, no doubt underlying. It has been suggested that this pigment, which causes "bleeding" on the shoulder area of the male and breast of the female, is due to the presence of the proposed "Ap" gene(s). Carefoot observed that carriers of the Recessive White gene in an inbred flock displayed Lemon Hackles. An example was his Partridge Exhibition male line where he consistently produced percentages of offspring which corresponded with Mendel's Law of segregation.
Classification: Recessive
Breeds True? √
Examples of Breeds with Recessive White: White Wyandottes, White Rocks, White Pekins.

Gold / Red (s+) - STURTEVANT (1912)

The Gold / Red genes are in all fowl unless the Dominant "Sex-Linked" Silver genes are present. Gold / Red pigment originates from the Red Jungle Fowl which represents the true natural appearance and distribution of Red pigment. Regardless of how many other pigment altering genes are present, if a fowl does not carry Silver, it is genetically Gold / Red in the respective areas such as Hackle, Shoulder, Wing Bay, and Sickles. However, these areas are not always identifiable, for example if a fowl is completely Black, it still has to be either genetically Gold or Silver.
Classification: Sex-Linked Recessive
Breeds True? √
Examples of Breeds with Gold / Red: Brown Leghorns, Welsummers, Partridge Wyandottes, Brown Red Old English Game, Wheaten Marans, Gold Dutch.

Thoughts for the Future

Looking back over the years of poultry breeding and how it has evolved, it would be very foolish to think we know it all at any given point in time. New developments are taking place continually, and no doubt there are many discoveries still to be made within the genetics of poultry plumage and characteristics in general.

Some traits may never be fully understood because they are "multi-factoral", or in scientific terms "polygenic" or "quantitative," meaning that more than one gene is responsible for the appearance of a single effect. One such example is the crest on many breeds. Some contributing genes are Dominant, and some are Recessive.

The best tools we have available to us are our eyes. Whilst genetic knowledge assists us greatly, it is selecting for utility traits as well as desirable aesthetic qualities that will help us not only to preserve, but also improve the quality of our breeding stock for future generations.

Best Wishes,

Grant Brereton, Ruthin, 2008

Glossary

Allele - one of two or more alternative forms of a gene at corresponding sites (loci) on chromosomes, which determine alternative characters in inheritance.

Aromatase - an enzyme whose function is to alter the conversion of testosterone, producing oestrogens.

Autosomal - genes that are on chromosomes other than the sex chromosomes, i.e. not sex linked.

Chromosome - one of the 39 pairs of genetic "sets of instructions" carried by a fowl, 1 from each pair inherited from each parent.

Clean shanks - legs (regardless of colour) that are clear of feathers.

Distal end - the end of the feather furthest from the body.

Dominant genes - genes which express fully in a single dose.

Epistasis – where a gene with a particular function is linked to another gene.

F1 - the result of crossing two birds that are believed to be unrelated. The term F2 is usually (and correctly) used to describe a sibling mating from the F1, however, it is often used by many in the hobby to describe offspring produced from mating the F1 back to the parent stock. F3, F4, F5 etc follow suit and are used by most hobbyists to describe each respective generational cross, whether to sibling, parent, or sometimes unrelated birds.

Gametes - sex cell, sperm or ova, containing just one of each pair of chromosomes that the adult bird carries. The term "Gametes" is most often referred to when a hybrid cross is produced, and used to predict the outcome of future crosses of a sibling mating through Punnett Squares.

Gene - the genetic coding of a trait.

Genotype - a fowl's actual genetic make up.

Incomplete Dominant - a gene whose effect is the result of being present in a single dose. Unlike Dominant genes, "Incomplete Dominant" genes produce a different effect when pure and so consequently do not breed true. A good example is the Blue gene.

Locus (plural loci) - the point(s) on a chromosome where a gene is carried.

Melanotic gene - a gene which is closely linked to the Pattern gene and whose basic function is to "Blacken" a fowl, but in combination with other genes, produces effects such as Spangling, Lacing and Double Lacing.

Phenotype - what a fowl visually appears to be.

Protoporphyrin - the organic chemical compound deposited in Brown pigment in eggshells.

Punnett Squares - a square divided into further squares (depending on the number of genes involved) to calculate what the offspring will carry in any given cross. Devised by the geneticist, Reginald Punnett.

Recessive Genes - genes which don't express in a single dose.

Tyrosinase negative – albinism, "Albino".

Tyrosinase positive - usually referred to in relation to partial albinism

Acknowledgements

First and foremost, I would like to thank my Mum and Dad, June and Alan for encouraging my interest in poultry from the age of six, and for all their help over the years.
My very patient girlfriend, Louise.
My uncle Jimmy and Auntie Ade for buying my first Light Sussex hens, which opened up my eyes to the world of poultry plumage.
Rob Boyd for all the help with supplying stock and for hatching and rearing many of my experiments over the years.
The Davies family, John, Pam, Richard and Nanette for allowing me to rear hundreds of birds on their property and for all their assistance in doing so.
The Bergeson family, in particular Andrew for making such good use of my cast offs as laying hens over the years.
Mario Griekspoor and his wife Wilma for putting me up on my frequent visits to Holland, and for all the help in sourcing quality birds.
Local gamekeeper Will Killow for his invaluable role in making use of my surplus cockerels for his ferrets, saving me the task of culling.
Ann Kendrick for allowing me access to many outbuildings and land on which to rear my birds for over a 4 year period.
Keith and Ann Lloyd for allowing me to keep my birds on their land, and for all their help on a daily basis.
On the genetics front, I have Brian Reeder to be most grateful to, for spending hours and hours responding to my inquisitive emails over a 6 year period and for being very patient with me.
Other genetic influences were: The late Dr Clive Carefoot, Clare Skelton, Ron Okimoto of the US, and David Hancox from Australia.

Thank you all very much!

Grant

Photo Acknowledgements

Paul Heath
Kong Vang
Clement Martin
Sigrid Van Dort
Louise Hidden
Geoff Parker
Toni-Marie Astin
Rob Boyd
Gareth Osborne
George Brown
Mr Adair
David Pownall
Matt Hanson
Ben Westby
Lucy Courtney
Rupert Stephenson
Jed Dwight